CONCILIUM

THEOLOGY IN THE AGE OF RENEWAL

CONCILIUM

CONCILIUM/VOL. 26

FUNDAMENTAL THEOLOGY

THE EVOLVING WORLD AND THEOLOGY

edited by JOHANNES METZ

VOLUME 26

CONCILIUM
theology in the age of renewal

PAULIST PRESS
NEW YORK, N.Y. / GLEN ROCK, N.J.

PAULIST PRESS
EXECUTIVE OFFICES: 304 W. 58th Street, New York, N.Y. and 21 Harristown Road, Glen Rock, N.J.
Executive Publisher: John A. Carr, C.S.P.
Executive Manager: Alvin A. Illig, C.S.P.
Asst. Executive Manager: Thomas E. Comber, C.S.P.

EDITORIAL OFFICES: 304 W. 58th Street, New York, N.Y.
Editor: Kevin A. Lynch, C.S.P.
Managing Editor: Urban P. Intondi

Printed and bound in the United States of America by
The Colonial Press Inc., Clinton, Mass.

CONTENTS

PART II

BIBLIOGRAPHICAL SURVEY

PART III

DO-C DOCUMENTATION CONCILIUM
Office of the Executive Secretary
Nijmegen, Netherlands

Preface

Johannes B. Metz/*Münster, W. Germany*
Werner Bröker/*Münster, W. Germany*
Willi Oelmüller/*Münster, W. Germany*

I

Evolution is the basic theme of this volume. It is viewed here principally as a border problem between theology and the physical sciences, although today the term "evolution" is also used in other fields, such as cultural and social history. One may well say that the problem of evolution brings out most acutely the tension between theology and science in their interpretation of the world. And theology is and will remain exposed to this tension. Although we know that faith is concerned with the whole of reality, theology cannot pretend to be a kind of universal science that can look down on the various sciences as merely derived or delegated functions of theology. Theology should quietly accept the fact that there are various kinds of knowledge and that it has to face this pluralism of knowledge constantly in the hope of achieving a fruitful exchange.

This pluralism will not allow theology to give way to the tendency of trying to create a superficial harmony and synthesis, somewhat in the sense of looking for "uncertainties" or gaps in scientific knowledge so that theology can then remedy the situation. On the other hand, theology cannot function as if there were a principle of "double truth" or take refuge in a pure fideism that would consider any knowledge acquired outside theology as simply irrelevant for the faith and so try to mark off, once and for all, a "neutral zone" for itself. There is no such

1

security. The tension remains and is, ultimately, the expression of the creaturely situation of the faith within a scientifically determined world. The bibliographical survey by Norbert Schiffers, on questions put by science to theology, confirms this.

II

We can deal with evolution at many levels. Therefore, a contribution on aspects of evolution (Bröker) presents the present position of the problem. Another article deals with the development of ecclesiastical statements regarding the theory of evolution (Alszeghy). A concrete example of this particular theological aspect of evolution is provided in the article on monogenism and evolution (Rahner), and this opens the theological discussion of the problem of evolution. This is continued with problems that touch the existential crux of the matter: sin in an evolutionary world (Dolch) and man's control of his own destiny (Dubarle). The article that should have dealt with evolution and immortality was not ready in time, partly because of the inherent difficulty of the matter. For this reason the office of the executive secretary of *Concilium* has provided some documentation on this topic.

The study of evolutionary progress and Christian promise (Cox) puts evolutionary thought into the context of an eschatological faith and also touches upon the basic question about the relationship between evolution and history. Originally we also had planned, in connection with this, an article on evolution as a "pattern" of future theological development, with reference to Teilhard de Chardin. Since the author fell ill, this article will be published in a later volume of *Concilium*. Lastly, the article on natural law and evolution (van Melsen) is concerned with a very urgent and related issue. The bibliographical survey by van Onna will give information about the contemporary debate on an important aspect of our general theme, namely, the original situation of man in the light of evolution.

III

Today the relation between theology and science obviously implies a multitude of general issues and problems. For example, the question of hermeneutics, i.e., how can one effectively talk about border problems of natural science in theology and pastoral care? Or, with regard to the connection between knowledge and sociology, how far are we influenced by the anonymous pressures of a monopolizing ideal concept of knowledge, namely, the domination of scientific and technical patterns of thought? How far do these scientific and technical patterns of thought rule our individual and collective consciousness? And if we look at our industrial society and our scientific civilization, how do we reconcile this technological mentality with that rational character and its compulsiveness which is the soul of our practical social life and our concrete historical activity, and for which this mentality has no means of expression? The answer cannot be found by directly confronting theology with science, but only by an *indirect* approach. It can only be discovered when the liberating and critical activity of theology aims directly at the *society* itself, a society dominated by this exclusively technological mentality. And so these scientific border problems increasingly demand a theology that is critically concerned with the formation of "awareness" in our technological society.

This general problem is given expression in various ways in the articles that deal with "the problem of the technical and scientific revolution" (Ellul) and "religious values in the age of technology" (Mesthene). The latter contribution looks from outside, as it were, at the critical situation and function of "the Churches" in our technological civilization. Together with the article on "scientific outlook and the Christian message" (Mascall) they try to explain, in a critical approach, what should be the starting point for theology and preaching in our technological age.

PART I
ARTICLES

Werner Bröker/*Münster, W. Germany*

Aspects of Evolution

For over a hundred years the thinking of Christian theologians has been concerned with the theory of evolution. At first, with a few exceptions, they saw the Christian doctrine of creation as threatened by the opposition that seemed to exist between the new hypothesis and the biblical account of creation. Hence, the debate between the theologians and the theorists of evolution was first limited to the subject of the bible and evolution. Today this problem has been settled to the extent that the Catholic Christian, without feeling obliged to justify himself in detail, is able to speak of the compatibility of the doctrines of evolution and of revelation, and beyond that, even accepts that the human body has derived from the world of animals. (The discussion concerning the unbroken transition from highly complicated, still lifeless material to living material hardly belongs within the framework of a theological debate.)

The cry is no longer "the creation story *or* the theory of evolution", but "the creation story *and* the theory of evolution". Both, the story and the theory, convey the fuller aspect of what is theologically called "creation". Peace has descended because positions have been surrendered that were occupied on account of prejudice and the doctrinal traditions of centuries.

But it would be a mistake to think that this "peaceful" solution that has come about in the last decades has done away with all the problems that the theory of evolution poses for theology. These no longer have the all-embracing importance of the relationship between evolutionary theory and creation story, but nevertheless they relate to vital points in Christian faith. These questions require answering all the more since the progress reports of evolutionary study are not only available in the special scientific journals, but have become accessible to the general public in a large number of popular science publications. We may assume that the idea of evolution, which today must be regarded as an essential part of the mental furniture of the Western world, will have an even greater influence on the thinking of future generations. For many, not least for the scientists themselves, the question of a world that is in the process of becoming will be one of the most important. Even today it is firmly posed by the published work of Teilhard de Chardin, which—whatever one thinks of the actual content of Teilhard's scheme—will perhaps stand in the history of thought as the research that expressed, for the whole world to consider, the latent questioning concerning the evolutive process of becoming. The number of theological essays that discuss Teilhard de Chardin's thesis in his terms shows how much theology also has been affected by the question of evolution.

If theology embarks on the discussion of questions that are posed for her by evolutionary theory, then it must observe one basic condition. It must not only hear the question just as it is formulated, but also attempt to understand the background of the question. This background has, it seems to me, two planes of differing dimensions. The first is the plane of the personal reaction of the questioner, who is impelled by his knowledge of the evolutionary process to ask what the significance of the process is; the second is the plane of the factual contents of the theory of evolution, which is the basis of all further questioning. This basis must be familiar to all those who concern themselves, whether theologically or scientifically, with the problems raised by evolutionary theory. But that means that an insight into the total perspective of the whole evolutionary phenomenon, as found in the appropriate sciences, must precede any theological comment. It would be to surrender any claim to full validity if one were to conceive the phenomenon of evolution in one's mind from the start in such a way that it corresponded to a desired or pre-given interpretation. It follows that the reasoned accounts of the evolutionary phenomenon given by the appropriate scientific disciplines are always, in their entirety, the criterion for theological comments also, and that even already accepted interpretations must constantly be overhauled when the appropriate sciences discover new aspects of their objects of research. Only in this way can one avoid vague, freewheeling discourse which often leads to any theological discussion of evolution not being taken seriously by "experts".

In the following pages I shall attempt to present what is understood today inside the natural sciences by the so-called "synthetic theory of evolution" or neo-Darwinism.[1]

I

THE PHENOMENA

(a) *Biological Evolution*

The discussion must start from the descriptions of the phenomenon of transformation given by the biological sciences. These sciences gave the idea of evolution the modern form it has today and filled it with the richest content. A full account of the scientific theory of the evolution of life would be divided into two partial theories, which together make up the theory of evolution in its modern form: the general theory of evolution and the specific theory of evolution. Dividing up the theory in this way has the advantage of making a more exact distinction between the part of evolutionary theory that has already passed into

[1] Recent descriptions of the phenomenon of evolution, which are relatively easy for the non-specialist to obtain, dispense me from a detailed account. May I mention especially P. Overhage, *Die Evolution des Lebendigen. Das Phänomen* (Quaestiones Disputatae 20/21) (Freiburg im Br., 1964).

scientific thought as a fixed and largely undiscussed contribution, namely, the phenomenon of evolution, and what is still in part open to the formation of hypotheses, namely, the (scientifically understood) causality of the evolutionary process. Not enough is known yet about this causality for it to be a basis for theological discussion—if, indeed, it appears necessary at all for such special areas.

Now the evolutionary phenomenon, as it is seen by the theorists of evolution, can be made clear to the understanding of the non-expert in two ways. First, one can show step by step what concrete biological or cosmic or historical form (morphological and functional) follows the other in time and space. In spite of the advantage of greater clarity, this kind of account has the drawback of presenting a vast amount of material; and if the account is not detailed enough, the suspicion that evolution is unilinear is hard to avoid. (And precisely that it is not unilinear is one of the main statements of modern evolutionary theory.) There is a second kind of account that does not have these disadvantages (and could avoid the danger of lack of clarity by producing as much illustrative material as desired). This account presents the basic ways in which the evolutive principle behaves, as worked out by the theoreticians of evolution in relation to the evolution phenomenon, thus giving a greater understanding of the phenomenon and showing more clearly which aspects are important for the borderline problems. The following account seeks to describe the phenomenon of evolution in the latter manner (briefly and without examples).

One of the first and basic elements of biological evolution is the "blood" connection of all life, due to the universal generative process, which is the reason for the basic similarity of the structural and functional structures of all living things. This life, proceeding out of itself in temporal succession, had and has the general tendency to spread out and fill all reachable areas of the inhabitable environment, including those that are created by the process of expansion itself. In addition to making full use of the external possibilities of taking possession, of expansion in space, there is also the tendency, coupled with spatial expansion, to realize inner possibilities, what one might term "living out one-self to the full".

In outlining the second basic element of the evolutionary proc-ess, the tendency of outer and inner fulfillment, we have also described a third element, that of the alteration of ancestral forms in their subsequent generations to become new forms and modes of life. The process of development exhausts in animals and plants all the conceivable phylogenetic possibilities of change. Evolution has constantly experimented with variations of form, and this experimentation has resulted in the great variety of forms observable. This experimentation is also the reason for the many biologically acceptable "structural mistakes" which ap-pear in the organic realm.

However, the constant creation of new forms and attitudes does not proceed arbitrarily, but evinces "orientation phenom-ena", "trends", i.e., continuing and dominating tendencies. This phenomenon, once called by many scientists "orthogenesis", ap-pears more and more to be a very involved and complicated process that is by no means rectilinear and that does not proceed at a regular pace. A large number of such trends, in every direc-tion that is biologically possible, exist in this kind of multifarious space-time correlation with one another. A first analysis of this enormously complicated evolutionary process already provides a considerable number of essential rules that enable one to predict with great probability and in a very detailed way shorter or longer successions of events in the history of a genus.

This evolutionary process, which leads through the various trends to the profusion of the forms of life, is not reversible. The organisms do not, as a rule, return completely to some ancestral condition. These directed, irreversible, evolutionary processes are events that build up organic structures, preserve them, com-bine and integrate them, then break them down again or de-velop them further, and thus in countless directions have created the vast variety of different structures, organs and forms.

Within this "production of a vast plenitude of forms of the

most various organic creatures" [2] and of attitudes, there is undoubtedly another tendency "which expresses itself in the total process of the evolution of life and which leads, over and above the various stages of organization, to ever new and higher types",[3] a phenomenon that has been called "biological ascent".[4] In the world of living things there is a constant development from lower to higher beings, even apart from man. This fact of a biological ascent is, it is important to note, not a general phenomenon of the history of organisms that can be found without exception in all organic groups and lines of development. Innumerable groups of animals and plants have died out, others show a regression or degeneration, whereas others again remain almost unchanged at the same stage of organization, and some develop only more perfect adaptations within their structure. The two ideas "evolution" and "progress" cannot be automatically equated.[5] Criteria for this development to a higher structural level can be seen comparatively only in retrospect: e.g., from the highly developed placental mammal type of today—to which man also belongs—we can look back to the organization of prehistoric reptiles, amphibious creatures, fishes and the like.

In such a retrospective view, however, essential elements of biological ascent are revealed. These are: (1) an increasing differentiation, which leads ultimately to better temporal and spatial analyses of the environment; (2) an increasing integration toward morphological and psychological unities (the more richly the differentiation develops, the more do the parts become subordinated); (3) an increasing independence on the part of the organism from its environment, which is connected with increasing differentiation and increasing integration and which at its

[2] *Ibid.*, p. 192.
[3] *Ibid.*, p. 204.
[4] *Idem,* "Der 'Biologische Aufstieg'," in *Stimmen der Zeit* 170 (1961/62), pp. 418-29.
[5] G. G. Simpson, *Auf den Spuren des Lebens, Die Bedeutung der Evolution* (Berlin, 1957), p. 147; English translation: *The Meaning of Evolution* (New York, 1951).

highest levels generally leads to an increase in the individual autonomy of the organism and in a certain sense to a selection of the various possibilities of action, to a kind of capacity for judgment by the senses, which makes possible action that shows insight and intelligence. Man regards himself as the highest product of this progressively refined evolutionary selection, if only because he lives among the forms of life that for the most part show no tendency toward higher development. This brief account, then, is a description of the biological evolutionary phenomenon and thus of the general theory of evolution with which both neo-Darwinists and finalists will agree.

This description of the phenomenon of biological evolution can, for good reasons, no longer be questioned. Certainly, further biological research will make a more detailed description possible in the future, and here and there alterations will even have to be made to the picture, but with the greatest probability the total structure will have to remain unchanged. It is a scientific theory that undoubtedly has the same degree of likelihood as the well-fashioned theories of the historical sciences. It is up to someone who casts doubt on the general theory of evolution to prove the opposite. This could only be done with a body of material and with an alternative theory based on it that is able easily to embrace within its system all the phenomena included in the general theory of evolution.

The preceding points concerning the evolution of organisms have been an attempt to present, if only in outline, the general theory as recognized today. But nowadays the idea of evolution is no longer limited to the organic realm, but is much more widely conceived and includes in principle the whole cosmos, the whole of creation. The evolutionary idea tends to universalism, and the corresponding sciences themselves are more and more making processes related to biological evolution the object of their research. Beyond the processes in the organic realm there are primarily the processes in the cosmos and the processes of history.

(b) *Cosmic Evolution*

The cosmic process, before which men have stood fascinated for thousands of years and which they have sought to fathom and explain, is seen by the physical and astronomical sciences more and more as a process of becoming and not merely of a basically static and unchangeable regular cycle. But this process of becoming cannot be grasped in a generally recognized cosmological theory. The mathematical models "still seem so very provisional that one can hardly speak of a thought-out model of the evolution of the world. On the one hand, the mathematical difficulties are very great, and on the other there is a shortage of observation material".[6] However much these hypotheses themselves diverge in their general statements about the process in the cosmos and of the cosmos, the one fact is the basis of the more or less scientific speculation, hypotheses and theories, namely, the variability of the process pattern in the cosmos, based on a changing dynamics. "The cosmological theories attempt to understand the whole universe as a unified expansion of movement, dependent on the matter and energy content of the world that takes place within space and time, and to express the rhythm of this movement as far as possible in the language of mathematics." [7] Cosmic process is not static, but dynamic, and it involves change. In this sense, but only in this sense, one can speak with due scientific probability of a cosmic development. We have here a cosmic parallel to the above-named third element of organic evolution.

It is probable that within the constantly changing cosmic process there are developments that are subject to the criterion of growing complexity and that, as in the organic process, can be described as "trends" and as realizations of various possibilities. (One has only to think of the transition of plasmatic matter to the cooled condition in which atomic and molecular structures are formed, which then shows laws that atomic physics čan investigate.)

[6] J. Meurers, "Kosmogonie," in *Lex. Theol. u. Kirche* VI, p. 573.
[7] H. Vogt, *Das Sein in der Sicht des Naturforschers* (Berlin, 1964).

The question whether it is possible to speak of a single, rec-tilinear development to a higher level cannot be answered. Such a general tendency is conceivable, but it is subject to at least all the scientific criteria that concern this type group itself. But if no definite statement can be made concerning a single line of progress running through the whole cosmos, but rather one can speak of progress in the sense of increasing complexity and uniformity only within the subsidiary processes of the cosmos, one cannot then say that living matter is the peak of a single straight total line of progress. One could say, however, that or-ganic matter, with its high molecular albumen structures, is the culmination of at least one line of development which has taken place in our solar system. Just as man, the peak of biological ascent, is not the peak of the only ascent of organic life but the last member of one line of evolution among many others, so one can generally regard organic matter as one peak of cosmic evolu-tion which, like man in the organic sphere, surpasses all other known evolutionary stages of cosmic matter in complexity, in-tegration, etc.

Thus, in the basic pattern of the cosmic and organic realms there is a certain resemblance, namely, the existence of a dy-namic variability with a large number of individual trends which are to be qualified in general by the particular degree of com-plexity and unification that they have reached; this enables one to say that the evolutionary phenomenon does not extend in its basic structure only to the organic, but is already an essential de-termining element of the cosmic (and inorganic) process.

In connection with the history of the cosmos the second law of thermodynamics is always mentioned as signifying, in a brief and pregnant way, something such as "the second law is the law of the historical quality of nature".[8] On condition that this second law is valid for the cosmos as a whole, it would be, when com-bined with the dimension of time, a guarantee of the irreversibil-ity of the great cosmic events. Increasing entropy in the course

[8] C. F. Weizsäcker, *Die Geschichte der Natur* (Göttingen, [3]1956), p. 35.

of time, together with the simultaneously proceeding cosmic change and variation would then justify the idea of a great, irreversible cosmic process moving toward an "end".

Organic evolution is involved in this total evolution toward an end—granting that the law of entropy is true of the cosmos as a whole[9]—since organic development takes place as the consequence of one among many other cosmic developments. In any case, organic evolution shares in the fate of the whole cosmos, in its "ascent" and in its "descent". The fate of life on the particular planet earth is already tied to the much smaller horizon of the solar system, whose entropy increases with relative independence. Life on the earth will perish when equilibrium of this energy has been established.

(c) *Historical Evolution*

If one could regard cosmogenesis as the basis of biogenesis or biogenesis as a continuation within cosmogenesis, then the history of man would be a continuation within biogenesis, as "the form of becoming which is peculiar to, or corresponds to, man".[10]

The course of history is not "mere process", "mere succession" or even process in constant cycles, but a genuine, irreversible growth in time, which again and again in the course of millennia has brought forth new forms of human life and adaptation. There is a bewildering variety of things coming into being and passing away both in the micro-historical and in the macro-historical realm. But this process of birth and death has not taken

[9] R. C. Tolman, *Relativity, Thermodynamics and Cosmology* ([6]1962), p. 486, advises caution: "Indeed, it is difficult to escape the feeling that the time span for the phenomena of the universe might be most appropriately taken as extending from minus infinity in the past to plus infinity in the future. The classical thermodynamic arguments against such a view must certainly be somewhat modified in the light of the increased possibilities of behavior provided by relativistic thermodynamics, and would be subject to even more serious modifications if the principle of energy conservation should fail within the interior of stars as suggested possible by Bohr."

[10] J. B. Lotz, "Geschichtlichkeit und Ewigkeit," in *Scholastik* 29 (1954), p. 488.

place, nor does it take place, with complete irregularity, with an incidence that is statistically random, but in a structured way, in trends, in such a way that one can perceive almost always an historically limited path followed by a human group, i.e., one can see a shape in space and time. The image of history resembles in a surprising way the structures of biological evolution (which says nothing about the similarity or dissimilarity of the particular factors operating). Just as the biological theory of evolution has shown that the profusion of forms does not exist only in spatial extension today beside one another, but also after one another in time, so historical science does this for the rich variety of the forms that humanity has taken in history. Human history is filled with a dynamism that seeks to realize ever anew the potentialities of being human.

Whether there is a principle of progress within this fluid dynamic process is hard to say. It is clear that if one applies particular criteria to certain features of history, progress will be found. Such criteria would be: (1) the numerical spread of humanity over the whole earth; (2) the increasing complexity of the social structure; (3) the increasing unification of humanity through the medium of civilization; (4) the increasing spread and growth of knowledge and technical ability. A survey of the particular areas in which there is progress enables one to speak to some extent of a general progress within the totality of history, in spite of the thousands of wrong directions, failures, world wars, etc. We may speak of the history of man as a phenomenon that corresponds largely, in its basic structure, to the phenomenon of phylogenesis.

This ascent of man, in accordance with a biological development, should not, however, be simply identified with the progress of humanity, since this identification would ignore the element of human freedom and of human capacities that are connected with freedom or are its condition, which is a basic category of history. If we do not acknowledge human freedom and see man as totally enmeshed in the causal structure of biological and cosmic development, or if we banish freedom from an account of history on heuristic principles, as something that cannot be known, we shall find it enough to hope for a further relative ascent of humanity. But if we believe in freedom, we shall not reject the progress outlined within the history of man, and hence we shall also look optimistically toward the future, but recognize at the same time that history, in its central process, at that point where freedom operates, does not manifest any progress, or at least any that can be easily discerned. The criterion for progress in this central process would have to be an extensive and intensive increase in freedom and an increase in signs that show that man is free. But whether freedom has increased in the course of human history is very questionable, particularly as it is very hard to discover to what extent this *conditio humana* has ever been a determining element in history. The share of human freedom in the working of history cannot be analyzed, as scientific method is scarcely able to separate it from man's simultaneous involvement in the course of events. One can speak of freedom only in a general way and surmise that it is perhaps present in outstanding moments of history, but historical situations can never be clearly compared in respect of the degree of freedom realized in them, nor can one ever speak of more or less freedom in such a comparison and certainly not of an unbroken growth in the course of a development. We can state how "primitive" the civilization of primeval man was in relation to our own, but we cannot say how "primitive" his freedom was.

In conclusion, we can say that human history, inasmuch as it is the history of culture and of civilization, manifests a process of growth that leads to continually new forms and modes of human life and thus is like humanity "living itself out to the full". Over and above the profusion in time and space, this process of growth constantly manifests lines of progress if it is measured by particular parameters. These lines of progress are by no means always straight, but are the products of much groping, experimentation and error. Human history, as the ascending process of growth, resembles the process of cosmic and biological growth, since it has the same structure—i.e., is coming into being. In human history this process of becoming can be marked by human freedom, through which the neutral ascending process can be positively or negatively qualified. Whether this qualifying freedom is itself reciprocally subject to progress cannot be scientifically demonstrated.

What has been said so far can be summed up in the following way: evolution as a dynamic change in time is not only a phenomenon of life (of individuals, races and species), but is equally a phenomenon of the cosmos and of history. Change and temporality are the basic structures of all nature, and in such a way that both are not to be logically reduced to the other, but appear given with each other. However one might describe time today (in contrast to the ancient conception), whether as "experience time", "physical time" or "time of history", there are always given with it the uniqueness of the event, the irreversibility of the process and the constant changing to something new.

But this process, which is characterized primarily by irreversibility in time, does not lead to a chaotic conglomeration of forms and kinds of function in time and space that is subject only to static laws of distribution, but it rather manifests tendencies, "trends" that one can describe as developmental unities and that have both a temporal-dynamic and a spatial-static order. These developmental unities are characterized by the interaction of the laws that are peculiar (in the widest sense) to matter, including the laws that emerge only in the course of the complication of material structures, and through matter (again in the widest sense) as a medium. Through closer scientific analysis it is possible, after some preliminary general observations, to discover in these developmental unities rules of development that enable predictions to be made, according to the exactness of the analysis. The discernible lines of development are not necessarily in every case those of progress; tendencies of decline are just as frequently possible, and indeed found.

The idea of higher development itself is problematical, since it depends on criteria that man as such determines inasmuch as they are not *clearly* pre-given, like gravitation (apart from the "formulation"), but it is capable of being interpreted in different ways, according to what man regards as high or low. It is understandable that in this connection man regards himself as the most advanced creature and takes from himself the parameters of progress.

By means of criteria that are ultimately taken from the structure of man, the most different types of space-time developmental unities within the cosmos can be judged relative to man. In this situation a tendency to progress can be ascribed to different developmental unities, since they correspond to one or the other parameter or even several at once. But although these lines progress in some way relative to man, one cannot say that they progress *toward* man; rather, most of them have been and are divergent *from* man; they end long before man even appears, or lead to results that today are a long way from man and make up the vast richness of the forms and modes of inanimate and animate matter.

That man is the most highly developed thing among all that is inanimate and animate needs no proof, as the parameters for progress are all contained within his constitution and taken from it. Looking backward from man, we are able to see among the multifarious intertwinings of the paths of evolution one that can be followed like a red thread in the evolutionary labyrinth from the beginnings of the structure of the world as we know it down to the present, the end of which man today holds in his hand. (Let us point out again that a similar thread can be followed back for all forms of matter that are found today.) But we cannot say that this thread runs in a straight line. It goes round many corners and detours in the labyrinth, indicating the many failures and mistakes on the way to man.

Whether this thread, to continue with our image, is uniformly woven or whether it is tied together at a few places still cannot be said today with absolute certainty, although many signs indicate a line of development running right through, especially at the point of transition from the inorganic to the organic.

Evolution, multifariously interwoven and taking place at the many levels we have suggested, is today—man firmly included —recognized as a basic element of the total cosmos, so that one may rightly speak of the world as in process of becoming. When this world, living itself out through the tremendously varied processes of evolution, moves into history, the freedom of man enters it—providing that one acknowledges such a thing as freedom—although it is still uncertain whether it is subject to evolution itself.

II

The "Majesty" of the Process

If evolution is determined in this way and with this content, we must also note that humanly important questions that can be raised with respect to evolution are concerned with an element of the evolutionary process which cannot be easily enclosed in scientific terminology, but which is immediately apparent to anyone who has dealt with it with any seriousness, namely the *majesty* of the process. Figures can be given that generally transcend the imagination and indicate the enormous extent of the process, but this knowledge does not approach the amazement that seizes man whenever he imagines to himself, on the basis of the material that has been investigated, the mighty unity and closedness of this multifarious process of forms, although in most cases it can only be known in smaller or larger sections. This element of amazement in view of the majesty of evolution is—however unscientific it may sound—the actual motive that sets the questioning going and sustains it.

III

QUESTIONS

Evolution, characterized as spatial and temporal and proceeding in a way that evokes amazement, especially the evolution of organisms, has become a problem to man. In the face of this process man has lost all ambition to be the center of the cosmos, as he thought automatically to be the case before the beginning of modern science. Knowledge of the temporal and spatial extent of the cosmos makes the central position of man questionable. Seeming like a dethroned king of the universe, he resists seeing the universe as an appendage of the human earth. Thus, the relation of the cosmos to man and of man to the cosmos has become problematical; the great reality is no longer automatically conducted according to human demands that correspond to man's nature. The "world" lives a life of its own, over against which the reflective man of our day sees himself as alone with the great question: What is the point of the vast spatial extent and temporal depth of the cosmos? Is there meaning to this world or is everything ultimately submerged "in the night of total meaninglessness"? [11] As he looks at this newly discovered universe the believer finds that the first question of the catechism acquires a new depth. "Why is man on earth?" is today transposed into "Why is man in this universe that has existed for billions of years before him and billions of light-years away from him?" "Has the whole of creation a center of meaning; has its cosmic and historical motion a common goal?" [12] The theory of evolution answers the "how", but not necessarily this "why".

In the long run man cannot be satisfied with the knowledge that evolution is a process of becoming, that it takes time and that its duration is creative. He cannot recognize evolution as a fact, assert it "and then relax without attempting to understand

[11] K. Heim, *Weltschöpfung und Weltende* (Hamburg, 1952), p. 132.
[12] J. Bernhart, *Die unbeweinte Kreatur, Reflexionen über das Tier* (Munich, 1961), pp. 45f.

its nature, sense and goal".[13] The general question of why this cosmic whole exists cannot, in view of the whole historical movement of humanity, be avoided by man. Where is it going? What is it going to be? "Does it lie sufficiently and permanently in the sphere of phenomenal reality?"[14]

In dealing with these great questions the statements of theology must not ignore "the world view of contemporary man, in which 'evolution' is both a fact and an ubiquitous concept".[15] Theology today, however, is not concerned only with such men, but the results of research into evolution are themselves of direct importance for it. From these results there follow many questions, ranging from the problem of natural law, general anthropological statements such as monogenism, the body-soul "dualism", special theological and anthropological statements such as the doctrines of the original state of man, original sin and sin in general, to the Christian conception of the future of the world and the end of time.

If the idea of "nature" has changed from the conception of a static and hierarchical succession of stages to that of a dynamic continuum to which ever fresh opportunities are opened up, then we are faced with the question—and this question does not already assume the answer—whether we must not also attempt a new derivation of the idea of natural law.

If a special scientific investigation makes it more and more probable that new morphological and functional organic forms arise in communities of propagation (populations) and are governed by their collective laws, then perhaps we may ask whether humanity may not claim an exception, namely, monogenism.

Is it theologically and absolutely necessary to maintain the anthropological dualism of body and soul, and in such a way

[13] G. Thils, *Theologie der irdischen Wirklichkeiten* (Salzburg, 1955), p. 242.
[14] *Ibid.*, p. 270.
[15] K. Rahner, in his introduction to Overhage, *op. cit.*, p. 5.

that what constitutes the human, the spiritual dimension or the soul, must be expressly excluded from the evolutive process?

If we see man as embedded in the evolutive process, then may not the things that particularly distress man, such as illness, want, suffering and death, be understood as a bitter ingredient that belongs to the evolutive as such, even to pre-human development, and does not need to be explained by the acceptance of a primeval human fall that brought about this situation? What, in this connection, do we mean by paradise?

And how is the sin of recent man to be theologically determined, if this sin also is to be understood as an evil that belongs naturally to the evolutionary structural level reached in man?

What does theology answer if it is asked for its attitude to the future plans of man for this world, determined as he is to direct with his own hands the course of human evolution?

What conception of the end of time can theology offer, when it becomes ever more clear that the material world as such is not moving toward any destruction, but will continue to exist in unlimited duration, even if the various structures that have evolved will collapse and perhaps emerge anew?

This small catalogue of questions alone shows how intensively theology is concerned with the process of evolution and will continue to be so. Ultimately it is concerned with the interrelation of salvation and world history. And thus we may, in conclusion, make the point again that the theologian who seeks to deal with these questions requires definite knowledge of evolution as a basis.

Zoltán Alszeghy, S.J./ *Rome, Italy*

Development in the Doctrinal Formulation of the Church concerning the Theory of Evolution

I

EVOLUTION AND THE ECCLESIASTICAL MAGISTERIUM

Provincial Council of Cologne

The first encounter between the ecclesiastical magisterium and evolution took place in 1860 on the occasion of the provincial Council of Cologne. In proposing principles for a Christian anthropology, the Council characterized as "completely contrary to scripture and the faith" the opinion that man, insofar as his body is concerned, is derived from the spontaneous transformation of an inferior nature; it is also contrary to scripture to doubt that the entire human race has descended from the first man, Adam.[1]

Vatican Council I

Such a position naturally had an effect in the preparation for Vatican Council I. In May, 1869, during the meetings of the Theological Commission, a consultor noted that a theory was being publicized which regarded man as a product of the spontaneous evolution of matter.[2] Nor was the conciliar *aula* itself free from some allusion to such a theory.[3] However, the prepara-

---

[1] *Collectio Lacensis*, 5, 292.
[2] *Mansi*, 49, 697.
[3] *Mansi*, 50, 163-166; 51, 128.

25

tory work focused its attention on polygenism whose connection with evolution had not yet been made evident.

Indeed, both schemas on Catholic doctrine, which were successively presented to the Council, anticipated the definition of the descendance of all men from a single couple as a dogma of the faith.[4] As was repeatedly noted in the Acts, there was no opposition to such a projected definition;[5] only Bishop Verot, the original Bishop of Savannah—though he admitted the possibility and suitability of the definition—expressed amazement that the geological and ethnological reasons adduced by polygenists should be described in the exposition of the schema as possessing little weight.[6] As we know, the interruption of the Council prevented an in-depth discussion of this question.

After Vatican Council I

Immediately after the Council, attempts began to be made to distinguish materialistic transformism from a theistic transformism which takes into account the essential difference between man and the other creatures of the material world. The theories of G. Miwart,[7] M. D. Leroy[8] and J. A. Zahm,[9] accepted in Italy by G. Bonomelli,[10] possess a common view that only the body would have been prepared by evolution; in this substratum God would then have infused the spiritual soul created by him.

Although Miwart's work drew no reaction from Rome, Leroy, Zahm and Bonomelli were advised that their opinions had been judged untenable "by the competent authority", since they were

[4] *Mansi*, 50, 70; 53, 170; *ibid.*, 175.
[5] *Mansi*, 53, 212; *ibid.*, 297.
[6] *Mansi*, 50, 108; *ibid.*, 165.
[7] *On the Genesis of Species* (London, 1871); *Lessons from Nature* (London, 1876).
[8] *L'évolution restreint aux espèces organiques* (Paris-Lyon, 1891); cf. *Revue Thomiste* 1 (1893), pp. 532-5.
[9] *Evolution and Dogma* (Chicago, 1896); *Evoluzione e dogma* (Siena, 1896).
[10] Cf. *La Civiltà cattolica* 17, Vol. 4 (1888), pp. 362-3.

contrary to sacred scripture and "sound philosophy".[11] Later, official declarations indicated that the "competent authority" was the Holy Office.[12]

This entire matter involves no action on the part of the ecclesiastical magisterium: in fact no document was published, and no one of the works in question was ever placed on the Index of Forbidden Books. The interventions against Leroy and Zahm are therefore merely disciplinary actions intended to prevent propagation among the faithful of opinions that were at the time commonly regarded as dangerous to faith. Indeed, the official comments accompanying these actions stressed the scientific inadequacy of evolution, and alluded to the possibility of a reexamination of the question "when evolution should have passed its examination at the tribunal of science".[13]

The 1909 Replies of the Biblical Commission

The year 1909 witnessed the appearance of the first document directed to the entire Church, which dealt with the problem of hominization. A series of replies of the Biblical Commission concerning the historical character of the first chapters of Genesis lists among "the facts narrated in these chapters, which touch the foundations of the Christian religion" and whose literal historical sense cannot be called into question, "the special creation of man, the formation of the first woman from the first man, the unity of the human race" (DS. 3514).

In speaking of the unity of mankind, the Biblical Commission unquestionably meant a strictly monogenistic unity; and it is at least difficult to suppose that in speaking of the origin of the woman the Commission did not intend to affirm a physical derivation from men. The expression *peculiaris creatio hominis* is less clear; it has commonly been interpreted as referring to a

[11] *Le monde* (March 4, 1895); *Civ. catt.* 17, Vol. 5 (1889), pp. 48-9; *ibid.*, Vol. 7 (1899), p. 125.
[12] *Civ. catt.* 18, Vol. 6 (1902), pp. 76-7.
[13] *Civ. catt.* 16, Vol. 9 (1897), pp. 202-3.

special action of God having as its object not only the soul but also the body of the first man; however, it does not appear that an evolution that allows for such an intervention is to be excluded.[14]

In any case, this document was in fact not in favor of an evolutionistic explanation of hominization. The assertion that special interventions of God interrupted the series of created causes in order to fashion the body of the first man and that of the first woman, results in isolating the origin of the human race from the context of total evolution, thus destroying its immanent intelligibility which constitutes the principal attraction of an evolutionistic view of the world. Nevertheless, as was authoritatively stated in 1948 (DS. 3862), the reply of the Biblical Commission did not exclude further research.

Type of Assent Required for Non-Infallible Official Documents

Indeed, according to the more popular explanation among theologians of the period, infallible documents declare the truth or falsity of an assertion, and consequently require an internal assent of the theoretical order. On the other hand, non-infallible documents—such as the replies of the Biblical Commission— refer to the "safe" character of a teaching; they judge whether such a teaching can be affirmed within a determined cultural context without involving the danger of placing some truth of the faith in doubt.

Thus, the value of the latter type of documents lies more on the practical level; they indicate what must be the intellectual attitude of the faithful in order not to expose their faith to danger. Accordingly, they oblige an internal assent (cf. DS. 2880; 3503), but only as long as it is possible to prefer one opinion to another and as long as the cultural context to which they refer perdures.[15]

---

[14] L. Janssens, Secretary of the Commission, citing the reply in his *Summa Theologica*, 7, 675, does not exclude an evolution that admits a special divine intervention in regard to the formation of the body.

[15] Cf. for example, J. Franzelin, *De traditione*, Thesis 12; L. Billot, *Tractatus de ecclesia*, Thesis 19.

The cultural context in which the 1909 document was promulgated underwent progressive change. Symptomatic of this change was the fact that an increasing number of theologians began to express the view that a theistic evolution was reconcilable with the faith—without provoking the intervention of the ecclesiastical authority. At the same time, those who opposed evolution themselves reduced the theological value of their opposition in succeeding editions of their works.[16]

Allocution of Pius XII

This led to a new intervention on the part of the magisterium, which took account of the changing situation: the allocution of Pius XII to the *Accademia Pontificia delle Scienze* on November 30, 1941.[17] In his talk the pope affirmed the essential difference between the animal world and mankind, which excludes the possibility that man can come from an inferior creature by way of a true generation. However, the strong emphasis placed on the absence of certain and definitive knowledge in the matter opened up a narrow path to further doctrinal development.[18] Actually, we can conceive of a physical link of descendency between the first man and an inferior creature which would not be a true generation. In such a case, God would essentially modify the operation of natural generation, producing an organism that would thus not be properly a "son" of the organism from which it stems. Still in the same talk, however, the pope declares with a certain insistence that the woman has come forth from man.

Humani Generis

The encyclical *Humani generis* (DS. 3895-3897) represents a new and forward step toward the reconciliation of evolution and the teaching of the Church. This document no longer speaks

[16] Regarding this development, cf. E. C. Messenger, *Evolution and Theology* (London, 1931); *ibid., Theology and Evolution* (London, 1949).

[17] *A.A.S.* 33 (1941), pp. 506-7.

[18] Cf. the interpretation of A. Bea in *Biblica* 25 (1944), p. 77.

of the woman originating from man; it makes the positive statement that discussion on the origin of the human body from a preexistent and living matter is permissible. But it adds that the problem is not resolved and revelation requires great moderation and caution in this matter; hence, it is necessary to be disposed to accept the eventual decisions of the ecclesiastical magisterium on this point.

Nevertheless, the encyclical teaches that Catholics are not free to accept polygenism. It does not give a direct revelation of the monogenistic origin of the human race as an argument, but it does assert the impossibility of seeing how polygenism can be reconciled with the dogma of original sin.

Documents after Pius XII

Documents after Pius XII touch only indirectly on the problem of evolution. Although taking account of the possibility of hominization through evolution,[19] they nonetheless affirm the necessity of proceeding with moderation, and they insist on the fact that the question of the reconciliation of the faith with evolution cannot yet be regarded as definitively resolved.[20]

A recent allocution of Paul VI to a group of theologians[21] characterizes evolution as no longer an hypothesis but a "theory", and makes no other reservation for its application to man than the immediate creation of each and every human soul and the decisive importance exerted on the lot of humanity by the disobedience of Adam, the "universal protoparent". In this connection, the pope observes that polygenism has not been scientifically demonstrated and cannot be admitted if it involves the denial of the dogma of original sin.[22]

[19] *A.A.S.* 44 (1952), p. 870.
[20] *A.A.S.* 42 (1950), p. 839; *ibid.,* 45 (1953), p. 604.
[21] *L'Osservatore Romano* (July 16, 1966), p. 1.
[22] The allocution received diverse interpretations: cf., for example, P. Dubarle in *Le monde* (August 6, 1966), p. 8; R. Rouquette in *Etudes* (1966), pp. 381-91; L. J. Lefèvre in *La pensée catholique* 102 (1966), pp. 29-37.

II

FACTORS INVOLVED IN THE CHANGED ATTITUDE
TOWARD EVOLUTION

This series of documents[23] does not exhibit so radical a change as has taken place for example in the matter of the authenticity of the "Johannine comma" (DS. 3681-3682); nor is the opinion acceptable in our day—regarded as erroneous by the Synod of Cologne—which holds that inferior living beings developed into man "spontaneously" in virtue of their nature. Still there is a change, insofar as the evolutionistic vision of the world is concerned. In the light of the most ancient documents of the magisterium, such a view could have appeared dangerous to the faith, whereas it no longer seems so today—although it is admitted that evolution is guided by God and that man is essentially different from other animals.

Change in the Manner of Presenting Evolution

This change is explained first of all by noting the manner in which evolution itself was presented. One hundred years ago evolution was the instrument of atheistic and materialistic propaganda; hence, to accept evolution in the form of that time really constituted a danger to the faith. Furthermore, there is little doubt that the arguments for evolution are more convincing nowadays. Theological opinion, which was for a long time inspired predominantly by the Aristotelian epistemology, had little regard for hypotheses; today, evolution as well as all theories of modern science proceed precisely through the verification of hypotheses.

Deeper Understanding of the Autonomy of the Sciences

Such verifications in practice do not signify the absolutely certain exclusion of every other explanation of the facts (for

[23] For a more extensive exposition, cf. J. A. De Aldama, "El evolucionismo antropológico ante el magistero de la iglesia," in *El evolucionismo en Filosofía y Teología* (Barcelona, 1956), pp. 237-52; M. Flick and Z. Alszeghy, *Il creatore* (Firenze, ³1964), pp. 274-303.

example, miraculous intervention on the part of superior beings). Therefore, since it was not confronted with a system based on evident principles, logical deductions and positive facts,[24] Catholic thought tended to explain the antinomy between the "obvious" interpretation of Genesis and the scientific theories along the lines of the contrast between dogma and "fallacious opinions, interpreted as if they were dictates of reason" (DS. 3017). A deeper understanding of the legitimate autonomy of the sciences[25] is more inclined to appreciate the value of theories based on the purely scientific verification of hypotheses.

Progress Made by Exegesis

However, it was chiefly the progress of exegesis that effected a change in the manner of interpreting Genesis 1-3 that proved decisive in leading ecclesiastical authorities to a less unfavorable attitude toward evolution. At the turn of the century the tendency prevailed to seek a historical nucleus in every part of the account of the origins, even if this were invested with poetic form; in addition, it was also thought that respect for the Word of God required a literal historical interpretation as long as this was not excluded with certainty.

These exegetical principles could not fail to produce an instinctive aversion to evolution. Recognition of the sapiential character of the Genesis account and its widespread explanation as a "biblical etiology" enables us to avoid searching the Word of God for the revelation of a scientific theory on the origins of living beings.

Deeper Understanding of the Creator's Special Action in the Creation of Man

A final factor that was to attenuate the diffidence of the Church toward evolution consisted in the deeper understanding of the creator's special action in the formation of man. For, on the one hand, it is inadmissible that the human race should

[24] Cf. *Civ. catt.* 16, Vol. 9 (1897), pp. 202-3.
[25] Cf. *Constitution on the Church in the Modern World*, n. 36.

spring forth independently of the creator; and on the other hand, the interpretation of the divine intervention in a determinative manner—as an action of God which is part of the same plane of secondary causes—does not fit in with an evolutionistic vision of the world. This obstacle has been overcome by conceiving the special action of God as one that works through all the generations of living beings, so that everyone shares in this special but continuous action in the great work of universal evolution.[26]

If we consider things in the abstract, further intervention on the part of the ecclesiastical magisterium could declare that evolution is irreconcilable with the faith. However, such a possibility would seem to be ruled out in the concrete. For the theological objections against evolution have already been abundantly developed; and if in spite of this fact the ecclesiastical magisterium has continued to allow freedom in the matter, such arguments are evidently unconvincing; neither can we reasonably expect other stronger arguments to be brought forth in the future.

[26] Cf. J. de Finance, *Existence et liberté* (Paris, 1955), pp. 258-66; P. Overhage and K. Rahner, *Das Problem der Hominisation* (Freiburg, 1961); M. Flick and Z. Alszeghy, *Il creatore* (Firenze, ³1964), pp. 285-8.

Harvey Cox/*Cambridge, Mass.*

Evolutionary Progress and Christian Promise

D
ifferent historical periods have
their love affairs with different
aspects of Christian doctrine.
During the decades just following World War I, Protestants
seemed nearly obsessed with the hiddenness of God and the sinfulness of man. In very recent years Roman Catholics have discovered anew the authentic universality of the Church and the
variety of charismatic gifts. In the coming decade, however, it
will certainly be eschatology, our understanding of Christian
promise, which will require the application of the best theological thought.

Our recent theological advances have been mainly in the fields
of christology and ecclesiology. However, they have required a
thorough reexamination not just of those doctrines but of all
doctrines, so interdependent is Christian theological thinking.
Similarly our new thinking in eschatology will require a relent-
less reappraisal of all our other doctrines, for eschatology is not
just one item on the agenda of theological deliberation; it provides the perspective from which all else must be understood.

But why do we need a more refined and biblically based
eschatology? There are two answers to this question.

The first reason why we need more theological work on eschatology is that our present doctrines in this field represent an
admixture of biblical and quasi-biblical concepts as well as

notions absorbed from the pagan cultural environment during
the history of Christian doctrinal development. We need to look
again at the biblical sources themselves, recognize where cultural
additives have distorted biblical eschatology and spell out the
characteristic contribution of the bible itself to our understanding of man's future. As we will show below this will necessitate
separating the prophetic, from the apocalyptic and teleological
strands.

Secondly we need a clear statement of Christian eschatology
because although the problem of the future and how to understand it has become the major interest of late 20th-century man,
yet our Western intellectual heritage has not provided us with
adequate categories for doing so. Theories of evolutionary progress, recently in intellectual disrepute, are becoming popular
again, not only in social philosophy but also in theology. This
interest in evolution springs not so much from modern man's
curiosity about his past as it does from his concern about his
future. Man looks at his origins to get some clue to his destiny.
Yet, as we shall show later on, there are distinct weaknesses in
evolutionary categories and in the teleological model on which
they are based. Only the rediscovery of a vital Christian eschatology will be able to confirm the valuable elements of evolutionary thinking while exposing its shortcomings and supplying
a different perspective.

I

APOCALYPTIC, TELEOLOGICAL AND PROPHETIC STRANDS

Hebrew religion had reached a kind of impasse in its eschatology just before the beginning of the Christian era. Some rabbis
taught that the messianic era would come on earth and in history.
Others, influenced by Persian motifs, believed that it would
come only in a blazing end to the present historical era. Christianity inherited from Israel both of these eschatological traditions
and never combined them with complete success. At times the

New Testament seems to say that the coming kingdom of God will transform and renew this earth. At other times this earth seems to be swallowed up in flames while a whole new and pristine world appears.

The orthodox doctrine of the Trinity, which taught that the God who had created the world and the God who was renewing it were one and the same, mitigated against a purely negative, antiworldly apocalypticism. So did the decision of the early Churches to retain the Hebrew scriptures, with their blatant earthiness, as an integral element of the bible. Christian theology thus escaped the temptation of becoming one more world-denying cult.

But no sooner had this battle been won when Christianity was faced with a new crisis from which it has still not sufficiently recovered. When it moved from Palestine into the Hellenistic culture of the Mediterranean basin, it had to adapt itself to the prevailing thought system of its day or be written off as just another provincial Jewish sect. Christian theologians did so by embracing Hellenistic philosophy, including a *teleological* view of history. The resulting mixture of Hellenism and Hebrewism provided the intellectual basis for the entire history of the West. God became the *ens,* being itself, and his attributes were those of changelessness, aseity and eternality. Centuries of theological battles were fought over how Jesus could have been divine, yet still have lived in history, suffered pain and died. Elaborate schemes about two natures united in one Person were thought out to answer this problem. Even today, we still move within assumptions and modes of thought laid down in this astounding cultural synthesis.

Thus we inherit from Christianity three different, even contradictory, ways of perceiving the future. The *apocalyptic,* deriving from ancient near Eastern dualism, foresees imminent catastrophe, produces a negative evaluation of this world and often believes in an elite which will be snatched from the inferno when everything else dissolves. The *teleological,* derived mainly from the Greeks but adopted by Christianity, sees the future as the unwinding of a purpose inherent in the universe itself or in its primal stuff, the development of the world toward a fixed end. It provides the basis for philosophies of social evolution. The *prophetic* is the characteristically Hebrew notion of the future as the open field of human hope and responsibility. The Israelite prophets did not, as many popular misconceptions would have it, "foretell the future". They recalled Yahweh's promise as a way of calling the Israelites into moral action in the present.

During the long millennium in which a "politics of the future" was not the major preoccupation of the West, this unstable compound could stay together. Indeed at those points where there were moments of intense fascination with the future—Joachim of Fiore (1132-1202), millinarian and chiliastic groups—the synthesis was threatened. Today, however, orientation toward the future is not a minority preoccupation but the mood of the whole culture. Hence the contradictions in the tradition become more noticeable, hastening the dissolution of the compound.

Western culture is now undergoing a period of rapid and extensive secularization. In such a period a society continues to rely on values and perspectives produced by an earlier, more religious period of its history, often without fully realizing it is doing so. Max Weber demonstrated how this could happen. He showed how attitudes implanted by Calvinist religious faith continued to motivate people after the explicit religious commitment had faded. Weber referred to "the ghost of a dead religious belief" and showed how early capitalist society was based on secularized versions of Calvinism.

Even today our civilization is guided by impulses and attitudes that are the secularized expressions of elements once expressly grounded in religious belief. This is especially true of our attitudes toward the future. We have just shown that Christianity was the heir to at least three separate strands of eschatological thinking which it combined more or less successfully: the apocalyptic, the teleological and the prophetic. Our various modern attitudes toward the future are secularized versions of these types of eschatologies, subtypes of them and mixtures of them.

A brief examination of the various types of secular eschatology abroad today confirms this assertion.

II

MODERN ATTITUDES TOWARD THE FUTURE

Classical apocalyptic imagery included both a vision of catastrophe and holocaust (the *dies irae*) and a celebration of the restored and glorious new world it would usher in. As the American literary critic R. W. B. Lewis has shown in his chapter "Days of Wrath and Laughter" in *Trials of the Word* (New Haven and London: Yale University Press, 1965), succeeding generations of secularization have evidenced a tendency to separate these two components. They have become preoccupied either with the new age itself, to the exclusion of any real interest in the catastrophe, or they have become fixated on the vials of fire and have forgotten about the new earth their purifying flames are supposed to bring about.

The first type of truncated apocalypticism toasts the coming new age with little reference to the preparatory catharsis and purgation. It arose mainly in the surge of hopefulness accompanying the initial successes of the American and French Revolutions. In his essay entitled "Prophecy, Apocalyptic and the Historical Hour" in *Pointing the Way* (New York: 1957), Martin Buber calls it "inverted apocalyptic". In its most optimistic form this non-catastrophic apocalypticism appears in the poetry of those English romantics who wrote in the period just after the French Revolution. Characteristically combining pagan and biblical elements, for example, Wordsworth writes in *The Excursion* (quoted by Lewis, *op. cit.* p. 200):

> I sang Saturnian rule
> Returned—a progeny of golden years
> Permitted to descend and bless mankind.
> With promises the Hebrew scriptures teem.

Although the Marxists preserved an awareness of the revolutionary catastrophe that stood between the present and the coming of the "Saturnian rule", and Marx himself refused to pencil in the details of the classless society, still Buber is largely correct when he says that Marxism remains the best example of immanent dialectic of "inverted apocalyptic". As we shall later argue, "evolutionary progress" theories of the future derive mainly not from apocalyptic but from teleological views of the future. Still there are both apocalyptic and teleological (as well as prophetic) components in Marx's own complex thinking, and a type of revisionist Marxism has maintained its belief in evolutionary progress since the 19th century.

1. *Contemporary Apocalyptic Thinking*

But the romantic exhilaration with the apocalyptic possibilities of the French Revolution lasted only a few years. It soon gave way to less sanguine views of the future. In our own time we have seen the emergence of a second type of truncated apocalyptic perspective: the one which focuses grimly on the destruction and terror that is to come and either indicates little interest in the new age it inaugurates or disbelieves that any new age will come at all.

The grounds of the appearance in our time of this negative apocalypticism are not hard to find. The glowing hopes for evolutionary progress with which our century was introduced were staggered by Verdun, Auschwitz and Hiroshima. A whole generation grew up with the nearly unbearable knowledge that a few old men in one of the capitals of the cold war empires could annihilate whole nations by pushing a button. The strident demands of the world's oppressed but impatient races and the coming of dozens of colonial lands to independence gave people in Western societies a feeling that their star had begun to set. Reaching back into the biblical imagery of the Negro preacher, and warning white people that the hour might already be too late to effect a reconciliation of the races, the black American writer James Baldwin used the following verselet in *The Fire Next Time* (New York, 1963) to conclude an essay:

God gave Noah the rainbow sign,
No more water, the fire next time!

But Baldwin is no pure apocalyptist. Just as the prophets once made use of apocalyptic and visionary rhetoric to call Israel to repentence, thus transforming these idioms into prophecy, so Baldwin and many like him are not understood with real accuracy if we simply classify them as apocalyptic. Baldwin mostly speaks as though it is far too late to avoid the ruinous consequences the centuries of racial injustice are bringing upon us today. But on another page he speaks as though there is still time, with the apocalyptic predictions serving mainly to heighten the urgency of the situation.

Still there is a style of political behavior that has arisen in recent years that might well be called the "politics of apocalypse". One expression of it is brilliantly described by Frank Kermode in an article entitled "The New Apocalyptists" in *Partisian Review* (Summer, 1966). Mr. Kermode shows that some of the great poets and writers in the English language in the 20th century have been apocalyptists. His examples are Ezra Pound, William Butler Yeats, T. S. Eliot and James Joyce. He then goes on to show that when the apocalyptic vision becomes more than a literary device or a fiction for organizing thought and goes on to provide the basis for a political order, then disastrous results ensue. Apocalyptic thinking breeds the fantasy of an elect who will be saved from the catastrophe and who may be called to rule the new world. In reminding us that Yeats, Pound and Eliot all had fascistic or ultra-conservative political leanings, Mr. Kermode comments that "the dreams of apocalypse, if they usurp waking thought, may be the worst dreams" (*ibid.*, p. 349).

Mr. Kermode is right in stating that when an elitist apocalyptic myth becomes the basis for a political movement the consequences are terrible, as we all saw with Fascism. But there is another and even more thoroughly nihilistic form of apocalyptic politics. It is the anti-politics of those groups who have lost all faith in earthly programs, who withdraw from genuine participation in the political order, and whose only interest in it is to harass and weaken it. Such groups appear on the fringes of both the left and the right today and become an additional hindrance to genuine social change.

Whether it is romantic, elitist or nihilistic, apocalyptic thinking is always politically destructive. Since Christianity has been the principal bearer of apocalyptic into the modern world, albeit an apocalyptic usually mixed with other mitigating elements, it is especially important that Christian theology today should purge and purify its eschatology. This will not in itself, however, rid us of the ghosts of previously religious views of the future which still prowl in our midst. To expose these persistent demons, theologians must also criticize the secularized forms of distorted eschatology as they appear in other cultural and political manifestations.

2. *Contemporary Teleological Thinking*

If apocalypticism has proliferated in its secularized stage into several subtypes, so has teleology. In one variety of teleological thought, each stage of the world's evolution is a deterioration and descent from the last. Life begins in a golden age and gradually loses its virtue and attractiveness until it becomes totally devoid of all value and significance. The Carnot-Clausius law, the so-called "second law of thermodynamics", when it becomes a world view lends a certain validity to this notion. The world is seen as "running down", tending toward a state of inert equilibrium. In this view, motion slows and stops, colds blend into hots, and finally lifeless equivalence reigns. Though it resembles the negative apocalyptic view in the unwelcome condition it foresees, the "winding down" view of the future includes no fiery holocaust. The world ends, as T. S. Eliot expressed it, "not with a bang but with a whimper".

There is a second type of evolutionary view also growing out of the teleological model but differing from the first in an important respect. It sees the world and man evolving, but the evolution eventually proceeds toward a state which will be identical with that from which it began. The *telos* is the recapitulation of the *arche*. It was natural for such a theory to grow up in societies where men derived their view of history from the cycle of the seasons, the ebb and flow of the great waters and the revolution of the heavenly bodies. Despite an indisputable evolution from spring to summer to autumn, winter always comes and then spring again. As Father Walter Ong, S.J., points out in his essay "Evolution and Cyclicism in Our Time" (in his book *In The Human Grain* [New York: Macmillan, 1967]), we now know that there is no such dependable cyclical pattern. Not even the earth's course around the sun is that stable. "In the universe as we now know it," he writes, "there exists no real model or analogue for cyclicism. . . . It appears that cyclic theories of cosmic evolution and of history depend upon setting up such a model or construct, for which one can find no exact counterpart anywhere in the universe, and upon using this model, despite everything, as though it applied to reality" (*op. cit.*, pp. 3-4).

Whether or not Ong is right that there is no evidence for cyclicism in the visible universe, it is certainly true that the most popular form of evolutionism abroad today is a belief in some sort of evolutionary progress. It can be based in biology and point to a Great Intelligence moving toward still unthought of forms of life. It can even be more encompassing. The wide popularity of Pierre Lecomte du Noüy's *Human Destiny* (1947) and more recently of the work of Teilhard de Chardin, despite the many disclaimers entered against him by paleontologists, surely comes from their effort to locate the human phenomenon within a purposeful cosmic process in which man, far from being lost, is seen as the key to the next stage of cosmic evolution.

It is easy to see an element of *hubris* social evolutionism in the teleological model on which it is based. Man experiences himself as a purposeful creature. Unable to believe that the vast cosmos around him is devoid of such purpose, he projects onto it his own purposive style, and usually assigns himself a crucial place in the *telos* of the cosmos. He has a vague sense that there must be some purposeful intelligence somewhere, that the appearance of a whole universe of such enormous complexity could hardly have happened by mere chance. This gnawing mixture of hope and wishful thinking still persists in even some of the most secularized people.

Evolutionism, however, like apocalypticism, fails to provide the perspective on the future needed for politics today. True, one can derive a certain comfort from the conviction that reality is moving toward a predetermined telos. But when this attitude informs our planning, it can inhibit imagination and discourage radical new initiatives. Any teleological philosophy obscures the fact that history is radically open, has no predetermined end and will go only where man takes it and nowhere else.

The weakness of teleological thinking is that it puts undue emphasis on the "arche", the beginning. The telos is really the highest development of the arche. The whole oak tree is there in the acorn and has but to develop and grow. History should be the realm of radical freedom and responsibility. Teleology projects onto history a way of thinking derived from nature, which is the realm of development and necessity. If nihilistic antipolitics is modern secularized apocalypticism, then teleology is the "nature religion" of modern secular man. As a nature religion it makes man feel a little more at home in the bewildering cosmos, a little closer to the plants, stars and animals. But it has disadvantages. It obscures man's character as an historical creature, as an animal with memory and hope who knows that if he destroys his world he can no longer blame it on forces beyond his control.

The most exciting chapter in the history of religion is the

titanic struggle that went on between Hebrew prophetism and the nature religion of the Canaanites. It was a battle between two views of man. Was man enmeshed in nature and akin to its vitalities and powers, or was man an historical creature, called by a God who acted in historical events and who required him to take responsibility for himself and his world on the way to an open future? Prophetism won this battle not by falsely extricating man from nature, nor by including the historical within the natural. It won, in effect, by bringing nature into history. The Canaanite fertility festivals became occasions to celebrate Yahweh's promises for the future. In the cycle of planting and harvest there is no genuine future, so the victory of prophetism over Canaanite baalism, though it took place long ago, remains important today. It assures the survival of a perspective on the future without which both planning and politics would seem futile.

3. *The Prophetic Perspective*

Recent biblical scholarship rejects the notion that the advent of Israelite prophecy marked a radical new departure in Hebrew religion. Israelite faith had always been "promissory".

For many reasons prophecy has often been mistakenly confused with soothsaying or prediction. Nothing could be farther from the truth. The Hebrew prophets, it is true, often used the contemporary rhetorical devices of their time, including envisioning the future, in their prophetic utterances. But their purpose was entirely different as was their view of history. They talked of the future to get people to change their present behavior. They did so because they believed the future was not predetermined but could be changed. As Hastings' *Dictionary of the Bible* rightly says: "The first task of prophecy is to break the people's faith in the ritual stereotype of destiny" (p. 808). The Jewish scriptures make a careful distinction between the *roeh* (seers) and the *nabi* (prophets). A true seer, such as Tieresias in Sophocles' *Oedipus Rex,* can foretell only because the gods have already determined every man's future. The seer speaks not to elicit repentance and a new course of action, as the prophets did, but to warn someone that striving to evade his fate is futile.

The prophets talk about the future in terms of what Yahweh will do *unless his people change their ways.* Yahweh is free to change his mind. The future is *not* predetermined. All that is sure is that Yahweh, who has promised to persevere with his people, will not abandon them. We should not allow either the literary form of prophecy or its misuses by Christian fundamentalism as spurious proofs of the Christian message to divert us from its main impact: seeing history as the field of man's moral responsibility for the future.

Is there a moral ethos today which expresses prophetism, as we saw for apocalyptic and teleology? In contrast to apocalyptic, the prophetic mood has confidence in the worth of moral and political action. It visualizes the future of this world not as an inferno that ushers in some other world but as the only world we have and the one which man is unavoidably summoned to shape in accord with his hopes and memories. The prophetic mentality rejects the apocalyptic notion that this or that elect group can escape cosmic ruination or is destined to rule the rest of us. It sees all peoples inextricably intertwined in the future of the world.

Against the teleological view, the prophetic sees the *eschata* transforming the *arche,* rather than vice versa. It sees the future with its manifold possibilities undoing the determinative grip of the past, of the beginning. In contrast to most forms of teleology, prophecy defines man as principally historical rather than as natural. Without denying his kinship to the beasts it insists that his freedom to hope and remember, his capacity to take responsibility for the future, is not an accident but defines his very nature. But, most importantly, prophecy sees everything in the light of its possibilities for human use and celebration. Without rejecting the influence of historical continuities, it insists that our interest in history, if it is not merely antiquarian, arises from our orientation toward the future. We write and rewrite the past, we bring it to remembrance, because we have a mission in the future. The Israelite prophets called the past to memory not to divinize it but to remind people that the God of the berith (covenant) still expected things from them in the future.

At its worst, Christian eschatology blurs and dilutes the Jewish view of the future by allowing too large a place for apocalyptic ideas and by compromising with teleology. At its best, however, Christian theology makes unequivocally universal a hope for the future from which non-Jews had sometimes felt excluded. By emphasizing the prophetic perspective, Christian theology will be able to confirm the valid insight of teleology (that history is not static but moves in one direction) as well as the vivid urgency of the apocalyptic view, but do so without sacrificing the unique insight of biblical faith: that the future is radically open and undetermined and that God calls man to the responsible shaping of that future.

Andreas van Melsen/*Nijmegen, Netherlands*

Natural Law and Evolution

EVOLUTION AND OUR CHANGING CONCEPT OF MAN

Whether the science of evolution affects our concept of natural law implies two problems that have to be examined. The first is that evolution affects the concept of "nature", and this obviously affects the traditional notion of "natural law", since this was based on a pre-evolutionary notion of nature. The second is the suggestion, prompted by evolution, that this evolution did not cease at the *origin* of man. And if man evolves, should not natural law evolve, too?

But the question cannot be limited to these two problems. Natural law is concerned with man, and if, therefore, we want to know what natural science teaches us about man, whether directly or indirectly, we must see not only what this science teaches us about man as *object* of this science but also about man as *subject*. For it is man himself who practices this science, and in this study he acquires not only a better understanding of nature (and this implies himself as part of nature) but also of himself as the practicing subject of this science. He realizes by experience what the possibilities of his knowledge are and how they should be exploited. True, what science unfolds about man as *subject* is not itself a matter of natural science but rather of philosophy. Yet, this only increases the importance of such data, particularly in relation to the consequences for the concept of natural law that belongs to the field of philosophy.

It is interesting to note here that what natural science teaches about man as *object* comes remarkably close to what it teaches about man as *subject*. In *both* cases we are brought up sharply to the issue of *evolution*. In natural science man was first made to realize that his scientific activity is essentially progressive. It is worth pointing out that in the whole work of St. Thomas there never occurs the thought that questions which science in his days was unable to cope with could possibly find a solution in the future. He was obviously aware of development of thought, which had taken place in the past, but he thought that this development had more or less reached its apex with Aristotle. This not only explains the great authority attributed to Aristotle but also the fact that Thomas's science was permeated by the idea that the first principles of every science were already known, so that the weight of scientific effort lay totally in the drawing of the correct conclusions from these first principles.

This is not without important consequences for Thomas's view on natural law, and obviously not only for *his* view. The whole traditional teaching of natural law betrays a view of science that affects even the very notion of natural law. For this reason I stated above that our discussion cannot be limited to what science tells us about man as *object*. Everything is far too closely intertwined here in any case. For the same reason I also pointed out that the 19th-century scientific discovery of nature (including man as part of nature) as *not* a static factor led in the same direction as the scientific view of man as scientist, i.e., the man who practices this scientific discipline. The idea of natural evolution fitted in well with the idea of historical evolution. And although one should not confuse these two, they corroborate each other's findings. The result was that modern man developed a new view, not only of nature, but also of himself.

Nature, then, was not something fixed once for all; it was no doubt subject to certain laws but these laws did not fix the factual natural order. This was plain in natural evolution, but even more so in man's ability to intervene in the data of the natural order. Man came to see himself as the creature that had to bring out the hidden potentialities of nature. Science was not merely a kind of contemplative knowledge of nature but also the means for transforming it in such a way that it was better adjusted to the human condition. And so man's technical activity took over the task of guiding natural evolution. But this relationship between science and technology was a mutual relationship. Knowledge could only grow as it intervened in the process of nature. Progress in science was seen to be closely linked with scientific experiment. Theory could only be verified by effective intervention in the processes of nature. This was the only way in which man could prove that he had real control. But it could never be more than a provisional proof, and man should be constantly ready for new experimental data that might affect and correct earlier views, but in any case could always enrich and deepen these previous views.

The important point here is that this mutual correlation does not only hold for science and technology or other forms of experimental science and practice, but also for philosophy. We have already seen how important a change in man's understanding of himself, of nature and of his relations with nature, has been brought about by comparing what man at first thought about all this with what he saw in actual fact. This involved such things as the necessity of experimenting and the constant progress of knowledge and ability.

II

Consequences for the Notion of Natural Law

What has all this to do with our notion of natural law? To make this clear I return again to St. Thomas's concept of natural law. Natural law is concerned with the first principles of action. These first principles are rooted in human nature and the tendencies embodied in it. They lay down how a man *ought* to act in accordance with his *nature,* if he wants to respond to the intentions of the creator. What is typical of man is that he must work out his purpose, not by the blind urge of his nature, as is the case with the other creatures, but by realizing the goodness of these aims, appreciating them and thus fulfilling them by his free choice. So the basic principle of natural law is, for St. Thomas, that good should be done and evil be avoided (*Summa Theol.,* Ia IIae, q.94, a.2). Although he emphasizes the distinction between human nature and that of other creatures, his general concept of nature obviously colors his view of human nature. We find there several points already mentioned and, in particular, the *static* concept of *nature* with a corresponding static concept of science. Thus, it is striking how confident Thomas is about the mind's ability to seize the unchangeable essence and to express this in fixed concepts.

However, Thomas often uses the term nature with the meaning of essence, so that when he speaks of natural law this means first of all that which flows from the *essence* of man and is therefore given with this essence. Seen in this light natural law has then not necessarily anything to do with natural things and their laws. But this is not really what he means either. It is clear that, in his attempt at defining the content of natural law more closely, he explicitly appeals to the order of nature, with its inherent purposes. From man's point of view natural law includes: (1) what man has in common with all created things; (2) what he has more particularly in common with the animal world, and (3) what specifies him as a rational being (*Summa Theol.* I, II, q.94, a.2). So Thomas's use of the term "natural law" is somewhat ambiguous. On the one hand, it refers to the (unchangeable) essence of man, and on the other, to the (unchangeable) order of nature. But it would be unfair to Thomas to attribute this ambiguity wholly to the ambiguity in the use of the term "nature". Behind his preoccupation with natural order in view of discovering the moral content of natural law there lies another preoccupation: what is moral must be founded in reality in one way or another. In the knowledge of himself and of his position in the world of nature man must discover what is "moral". For it is in the reality of nature, with all its purposes, that the purpose of the creator lies expressed. So, for St. Thomas, natural law includes the fact that man is created.

One can maintain that in a certain sense the basic inspiration of this view of natural law is still valid, on condition, obviously, that we eliminate the consequences of the static concept with regard to nature, man and science. This does not mean in the least that these consequences are ultimately of less importance. They touch the root of the matter and force us to revise the whole concept, since we have learned to see the man-nature relationship in a different light than was formerly possible. Thus, the existing order of nature, as we experience it, has lost its normative character; if nature shows us God's purpose today, it is precisely insofar as it clearly shows that it is man's mission to seize hold of this order and to make it serve the development of man. In this view, the key to the moral order is now seen to lie in *man*'s being. This in itself is not a small matter, but by no means the most drastic consequence.

It can be put briefly as follows. On the one hand, the content of natural law must be discovered in man's being; on the other, this being develops in human activity which in turn must follow the lines of natural law. This looks like a vicious circle. Man's true being comes out only in the unfolding of this being, so natural law can only be known through the process of this unfolding; and yet, in this process man must be guided by this same natural law. This becomes still more difficult when we realize that this unfolding always depends on the mutual relationship between theory and practice. It is therefore theoretically impossible to determine beforehand in which direction man must develop in order to put it into practice afterward. Only practice can make clear in which direction development is possible. But does it then not follow that there is no point in talking about natural law as the body of unchangeable basic moral principles? For, if we can still talk of a human nature in the sense of man's being, this being is insufficiently known to discover there those

moral principles with any clear certainty. To say that, although
this being is not sufficiently known, it can nevertheless contain an
unchangeable nucleus that persists throughout all development,
is no way out. The question is precisely whether such expressions
as "unchangeable nucleus of being" or similar ones do not derive
their meaning and usage from that static concept of reality and
science, as when one says that reality does indeed change but
only accidentally, and science indeed develops but only by a
new application of the *same* principles though in other circumstances.

But a dynamic concept of man, reality and science, makes it
impossible to determine once and for all what belongs to this
nucleus and what not. Only development itself can tell us; development does not only mean a new application of the *same*
basic principles, but forces us to revise these principles themselves constantly. At least that is the present state of affairs in
science. Why should ethics be an exception to this rule? There
is indeed little reason to suppose this, although many treatises
that deal with natural law still proceed from the same concept
of science as St. Thomas. This is bound to have a disastrous effect, because it leads to a highly inadequate treatment of new
developments and the problems they bring along with them, or
to a rejection of the very idea of natural law, and this would
open the door to every kind of situation ethics and relativism.
Yet, there is a solution, and this demands that we consider the
development of natural law in the light of what scientific development tells us about the first principles of a science.

III

WHAT DOES THE DEVELOPMENT OF THE CONCEPT
OF SCIENCE TELL US ABOUT MAN'S BEING?

St. Thomas's idea that the first principles of science are
known is no longer tenable, at least not without some basic
modifications. It is no longer valid for the empirical sciences be-

cause there the principles must be found in scientific research itself and are constantly subject to revision. Nor does it hold any longer for modern mathematics, even though at first sight it still looks very much like the Aristotelian and Thomistic concept of science, with its axiomatic structure. For the axioms of a mathematical system do not pretend to give us a knowledge of reality but are drawn up more or less *ad libitum*. One or other set of axioms may well be applicable to a specific area of reality, for example, that of material nature, but in such a case it must be empirically established. What the situation is in philosophy, will be dealt with a little later.

But first I must refer again to empirical science. When I stated above that Thomas's idea of first principles had been overtaken by the development of science, this was only a half-truth. What Thomas called "first principles" has in fact broken up into two categories of principles, which must be clearly distinguished. One may illustrate this as follows. It is possible to say that it is a *basic principle* of natural science that experimental facts (and therefore not the theory) have the last word. But one can also say that the law of the conservation of energy is a *basic principle* of natural science. In both cases we rightly speak of a *basic principle,* yet, they are of an essentially different kind. That there is this conservation of energy is a principle discovered by natural science itself. It may have to be modified by new experimental data, and has already been modified in the course of history. Such a modification takes place wholly *within* the scope of natural science; this science remains what it was but its content is changed.

The other basic principle mentioned above is very different. That experimental facts have the last word in natural science does not determine the *content* of this science but establishes something about the *character* of natural science, the sort of science and its method. If this principle were thrown overboard, natural science would cease to be itself. In other words, this principle is a *constitutive* principle of natural science. The important point here is that such constitutive principles imply a specific view of human knowledge and at the same time a view of the object of natural science, namely, nature, and a view of the relation between man and nature. These views are *implicit* because natural science never explicitly examines them (*in actu signato*), but they are acted upon in practice because of the method of natural science (*in actu exercito*). The explicit treatment of these principles takes place in philosophy, as in fact happened in Greek philosophy which devised explicit views of man and nature in such a way that science could thrive upon them.

And yet, there is a vast difference between the Greek view and the modern one, as I have already pointed out. How, then, did this new concept of science develop? Not through an exclusively philosophical approach to the question of man's knowledge but through what was revealed in the actual practice of science. For this practice did not only discover new things about the *content* but also about the *method,* and this while remaining wholly faithful to the original intent of the Greek view which was that the goal of science was the disinterested pursuit of the truth. This pursuit showed that not only the senses but also actual "handling" had to be incorporated in the process in a far more intensive way than the Greeks thought was either due or possible. Thus, natural science grew from a mainly rational science by way of empirical science into an experimental science. In other words, the constitutive first principles of science underwent in the active practice of science a development that would never have taken place without this active practice. No logical analysis can ever bring out the original richness of what is implicitly contained in the principle; only that actual "practice" of the principles can show what they contain. These principles show a characterization of man's being, by no means less important for showing one aspect of man, namely, man as the "practitioner" of science.

What has just been said can be put in still another way: it is in the "practice", the "working out" of his being that man realizes more and more who and what he is. This leads to two important conclusions. The first is that philosophy, too, cannot simply deduce its theses from specific first principles. Since we have come to realize that human existence is a dynamic process, philosophy, no more than mathematics or empirical science, can be considered as a kind of science for which the Aristotelian and Thomistic concept of science would still be valid. In philosophy, too, the first principles are known implicitly rather than explicitly. This seems to conflict with the "essence" of philosophy whose task it is precisely to make explicit what is implicit. What good then is implicit knowledge to philosophy? Would it not be simpler to say that, as an attempt to pin down man's being in the shape of definite concepts, philosophy is finished? No, because, although it is true that man only discovers who and what he is in the "practice", the "working out" of his being, it also means that this "practice" is only possible if the initial insight into his own being, which gives birth to this process, is already in some way true to reality. Science can only develop there where man has already seen, however imperfectly, something of the true nature of science and of himself as orientated to this science. And this is valid in general. Only by reflecting upon himself can man see in which direction human existence should develop. These considerations can in no way be classified as relativism, even though they refuse to attach absolute significance to a particular phase of the explicitation of man's being, in the sense that this absolute significance presents us with a final and unchangeable statement on man's being. The process of hominization continues forever.

IV

CONSEQUENCES FOR ETHICS

One can also apply this line of thought to the first principles of ethics. It is the obvious thing to do particularly within an ethical framework that incorporates *natural law*. For to talk in terms of natural law always implies that the principles of morality are rooted in nature, that is, in man's being. If we apply what has been said about first principles to the moral order, this will mean that in the moral order, too, the first principles are operative implicitly rather than explicitly. Here, too, they contain implicitly far more than man in some period of history is fully conscious of and can express in some explicit formula. Every explicit formulation is branded by the mark of the respective period, of what, at that time, was seen as possible. When new possibilities arise in man's existence, which must be assessed morally, it is therefore very possible that the traditional moral formulation falls short of the reality. But this does not prevent man from seeing in these new developments possibilities for the realization of ethical values that he has always appreciated. This process can best be described as a new *realization,* which is at the same time a *re-cognition.* What is new can occasionally conflict with older formulas, and yet one recognizes there some moral value because one recognizes the new as something that was already implicitly contained in earlier principles, even though at that time the consequences were not understood.

An example may illustrate this. Formerly slavery was accepted as not a contradiction of human dignity, and as a consequence of the "natural order". Yet, the rejection of slavery does not mean that our view of man is totally different from that of older cultures where slavery was accepted. But the soul of the earlier view was encumbered by many factors which we now see as time-conditioned and by no means essential although man did not see them like that in earlier days. This leads inevitably to the conclusion that *our* interpretation of man and *our* understanding of natural law are subject to the same development in the future. Today's controversies about morality show in any case how little clarity there is in our own views. And in this connection I would like to end with two observations.

The first is that there should not be such a negative attitude toward the fact that there *are* controversies in the field of morality. They simply reflect the dynamic situation of man, who realizes that he has not yet reached the end and who wants to think and act ahead in many directions. Formerly this division in morality as in philosophy could well be considered a "scandal", and it was attributed merely to a lack of straight thinking or—what was worse—to perverse intentions. Hence the dogmatic way in which each defended his own point of view. Today we may hope that what is not yet clear, may find some solution in the future. The past shows enough examples to justify some faith in this process.

The second point is particularly relevant in a theological publication. When morality is saddled with so many uncertainties, even where the explicit knowledge of natural law is concerned, there is a great temptation to crush this uncertainty by an appeal to revelation and to decide what is natural law and what not in this way. It seems to me that the interpretation of what it means to be "man" is, *mutatis mutandis,* subject to the same kind of limitations as I set out above with regard to the moral interpretation of natural law. The interpretation in the faith also knows an evolution, and it, too, is unable to force the process. Attempts to force it have done immense damage to Christianity in recent times and have at the same time discredited the very idea of natural law. And this happened not so much by attributing to natural law things that did not belong there, but rather by a lack of understanding of the way natural law functions as a comprehensive summary of moral first principles. For this reason in this short article on natural law and evolution[1] I have concentrated rather on the changed view of the concept of natural law itself as suggested by evolution (in the broad sense) than on concrete details of the content that are in need of modification.

[1] For a more detailed treatment of the moral problems, cf. my *Natuurwetenschap en Ethiek* (Antwerp, 1967); for the idea of evolution, cf. my *Evolutie en Wijsbegeerte* (Utrecht, 1964).

Karl Rahner, S.J./*Munich, W. Germany*

Evolution and Original Sin

I

THE STARTING POINT OF THE DISCUSSION

The question whether evolution is compatible with the teaching on original sin implies a further question: Does original sin imply monogenism? In other words, does it exclude polygenism or not? One can approach the problem either by trying to provide a direct proof for monogenism or indirectly by showing that it is necessarily implied in original sin as defined or is at least theologically binding. One can look for the direct proof either in the sources of revelation or in natural scientific knowledge. While the latter approach can never by itself produce a theological statement, it may nevertheless, if it is certain, lead to a possible reconciliation of traditional teaching about original sin with polygenism.

II

HOW FAR IS MONOGENISM A THEOLOGICAL CERTAINTY?

This question is the same as asking what theological value must be attached to monogenism. As such it can only be answered step by step.

1. *The Data of the Old Testament*

Monogenism cannot be proved directly by the Old Testament. If we interpret Genesis 1-3 correctly according to its literary genre and consider how man's (and therefore, the *first* man's) origin is revealed—that is, as a retrospective view of how man came about—the Old Testament tells us nothing about monogenism. This seems to be generally agreed among leading Old Testament scholars. Nor does the Old Testament provide an indirect proof. A genuine teaching about original sin, as distinct from an original deficiency, can only be understood if we presuppose that man was sanctified by the absolute *holy* Spirit of God before taking a personal moral decision; however, such a teaching does not occur in the Old Testament, nor could it occur. The mere fact of an original deficiency (such as death) does not prove monogenism, since such a deficiency can also be thought of without having recourse to a moral guilt and is not necessarily bound up with one single human being.

2. *The Data of the New Testament*

Insofar as direct statements are concerned, the New Testament repeats the Old Testament. *These* direct statements can therefore be interpreted without trouble along the same lines as those of the Old Testament. In Paul, however, the New Testament already contains a real teaching of original sin, so that there is already in essence an ecclesiastical teaching on original sin if we interpret Paul and the Council of Trent correctly.

The question, therefore, as to whether the New Testament teaching on original sin binds us to the view of monogenism can be reframed as follows: Does the teaching of the magisterium on original sin demand, clearly and without doubt, the acceptance of monogenism? But here we must first observe that when Paul in the New Testament uses the statement about one single first man, Adam, in his formulation of the teaching on original sin, we have to check *this* particular statement very carefully in order to see whether it means more in this context than elsewhere, or whether it merely repeats the formulation of the Old

Testament and therefore provides no clear teaching on monogenism.

3. *The Dogmatic Situation*

Monogenism is not a dogma defined by the Church. Theologians disagree about the exact qualification of this teaching, and when it was described as less important than a dogma there was no official condemnation. Strictly speaking, this fact is enough to prevent it from being a dogma. It is true that the Council of Trent presupposed an "Adam who is physically one". But in this case it was speaking of original sin and simply repeating the statement of scripture and tradition. It did not define monogenism since this question was neither put nor intended.

One can therefore only see an implicit and strictly binding teaching of monogenism as such in the decree of Trent if it is strictly proved that the Council's teaching on original sin as such cannot be held without this presupposition of monogenism. In any case there was no formal dogma of monogenism proclaimed at Trent. Moreover, the statements of *Humani generis* do not, in their only binding sense, assume that monogenism is a dogma of the extraordinary or ordinary magistrium, a teaching that is so certain and irrevocable that no theologian can consider a change in the theological position of the magisterium. Nor can we find any such irrevocable statement in the ordinary teaching of the Church. *Formally* and *today* the situation seems to me to be exactly the same as the evolutionary concept of hominization between 1850 and 1900. In those days one could not describe this concept as "free" and it appeared as if the magisterium allowed no change because of the unanimous rejection of the theory by the theologians.

4. *Can Monogenism Be Proved Philosophically?*

As long as the philosophical grounds for accepting monogenism are not certain and generally accepted, the theologian cannot dispense with the question as to whether polygenism is perhaps also compatible with the teaching on original sin. Such compatibility is at least thinkable, even if polygenism were, objectively speaking, false.

III
POSITION ACCORDING TO THE NATURAL SCIENCES

The great majority of scientists today accept the evolutionary process of hominization, as well as the theory of polygenism, if they still see this as a separate problem. This is understandable in view of their methodology. Natural science sees the advent of man as an event of the biosphere and its history. It does not consider the personal and spiritual aspect of man, at least not methodically and, in this sense, correctly. It has therefore no reason to consider hominization as happening only once in one single case, since biological events happen elsewhere in a number of cases of specifically the same kind. It is not the task of a theologian to provide the positive and *a posteriori* reasons that scientists have found in favor of polygenism. Nor is it his business to assess their importance. He can only state the fact that scientific anthropologists of today think in terms of polygenism.

IV
A THEORY PROPOSED

The real problem for the theologian lies in the question as to whether or not the teaching on original sin excludes polygenism.

1. *Thesis*

I want to try to prove the following thesis: In the present state of theology and science it cannot be proved that polygenism conflicts with orthodox teaching on original sin. It would therefore be better if the magisterium refrained from censuring polygenism.

In this argument I take for granted what has already been said, as well as the general methodological principles of theology. I also take for granted the traditional teaching on original sin, both in its defined form and in its generally accepted content, without bothering about new interpretations (as have been attempted by Schoonenberg, Haag and others). In this respect I shall strictly concentrate on monogenism as the issue under debate, that is, the question whether all human beings existing on this earth are, strictly speaking, descended from one single *human* couple by procreation. I shall not touch on other points of the traditional teaching.

The question then becomes: Is the free decision—required at the beginning of man's origin (*humanitas originans* as distinct from *humanitas originata*) to explain a failure in sanctity in mankind from its beginning as something that ought not to have happened—only thinkable as the decision of one single man (or one single couple) or by presupposing that mankind originated in only one couple? Or can the teaching on original sin also be maintained without this presupposition?

2. *Adam "and" Eve?*

The first question a theologian should seriously ask himself is: Can the Church logically and by the very nature of the matter leave us free to accept anthropological evolution on the one hand—as she does (DS 3896)—and on the other condemn polygenism? This is the situation:

(a) If evolutionary hominization is acceptable, we have to accept that "Eve" came about in the same way as "Adam". Any other view can only be a worthless compromise which is hardly to be encouraged. The decision of the Biblical Commission in 1909 about the "formation of the first woman out of the first man" (DS 3514) is no longer tenable in its exclusive literal sense if one accepts in general with Pius XII the evolutionary origin of man (which basically *conflicts* with this decree). We cannot think of "Adam" in terms of evolution and deny this for "Eve". This follows the thought of Pius XII's theologians. They argued that one cannot conclude to an evolutionary origin of Adam from the existence of Eve and from the statement given out by the Biblical Commission. But by the same token one cannot accept evolution for "Adam" and then reject it for "Eve". Polygenism can therefore no longer be rejected in the case of one couple.

(b) We can then rightly ask how one can explain in a convincing manner that the mutually independent origin of two human beings from the animal world must be limited to these two only. One may take refuge in every kind of *ad hoc* argument, such as the arbitrary will of the creator or the fact that hominization is in any case a rare occurrence on biological grounds, but do not such explanations sound rather forced? And then we have to ask ourselves further how one can understand that this one "Adam" found this one "Eve", both having evolved independently of each other, without appealing to miraculous interventions by God for which there is no justification. In other words, is it seriously probable that, within the wider population unit of the immediately preceding pre-hominids who create the living conditions and opportunity, only these two break through to become human beings and begin to procreate human beings?

3. *Polygenism and the Bodily, Historical Unity of Originating Man (Humanitas Originans)*

It is doubtful, to say the least, whether a bodily, historical *unity* of the first human beings can only be understood in terms of monogenism. It is a general principle of biology that true, *concrete* genetic unity is not found in the individual (but in the population within which alone many individuals can exist in the same biotype area in which the main environmental conditions and biotypes adapted to them are uniform) and with the same biotype (organisms of the same genetic constitution). Only within such a situation can evolution come about since selection can exercise its pressure only within such a population and not in isolated individuals. A population is always more, and as such more real, than the "species" of logic, which consists of many individuals in the abstract, through philosophical simplification, regardless of the question as to whether in fact the species has a monogenetic or polygenetic origin in the distant or not so distant past.

This biological-historical unity can also describe the state of mankind at its origin (*humanitas originans*) without limiting it to one single couple. Therefore, *mankind remains a biological-historical unity, even in terms of polygenism:* (a) because of the real unity (i.e., based on real factors and not on a process of abstraction) of physical existence in an ambiance which should not be taken as a mere empty "place"; (b) because of the real unity of the animal population from which mankind descended, and within which unity the pressure of selection toward evolution can be understood as real and leading to hominization; (c) because of the unity of the concrete biotype within which alone mankind can endure and procreate, regardless of the fact as to whether mankind started with one or more couples; (d) because of the actual human and personal intercommunication, which, in any case, is not merely a result but a constitutive element of the biological and *historical* unity of mankind as such (since "culture", speech, etc. belong to human "nature" as such and are not luxuries without which man could nevertheless exist biologically); (e) finally, because of the unity of man's destiny toward a supernatural aim and Christ, and this is not merely related to mankind as one but makes this oneness even more radically one. Something else has to be added.

We should not let ourselves slip into a kind of polyphylism, rejected by most anthropologists on scientific grounds; otherwise the biotype in which hominization (even a polygenetic hominization) took place would be divided into (biologically and humanly) completely disparate and independent areas. This holds good even though science does not yet know today *where exactly* on earth this hominization took place. It would also be wiser not to imagine this area as very small; otherwise one would only be embarrassed by future discoveries in paleontology.

Polygenism, therefore, allows us to imagine a hominization area where those beings that originated mankind formed a genuine biological and historical unit, achieved through a genuinely possible personal communication process. Such a regional limitation of the hominization area which made a real original unity of man's origins possible in the history of mankind would still be scientifically acceptable if we believe that mankind was prepared by a very *old* special line in the development of evolution.

4. *The Sin of One Couple or a Collective Sin?*

In view of what has been said, one can now ask whether in these conditions belief in original sin demands that this sin originated only with one couple and that only through this one couple mankind was deprived of sanctifying grace. This question, to which there is in our argument *no* definite answer, can be broken up into two distinct parts:

(a) Can one think of original sin, in a polygenetic but unified origin of mankind, as having started with one single man (or one couple) so that existing mankind was deprived of sanctification, and can one think of it in a way that neither eliminates the teaching on original sin nor makes it appear as arbitrary and for that reason *not* real?

(b) Can one think of original sin as the sin of the whole original group of human beings, polygenetically one, and that this whole group, historically united and the vanguard of mankind, committed collectively what is called *peccatum originale originans,* original sin at the start, and that this group can be personified as "Adam" because it represents a genuine unit?

5. *"Adam" Understood as a Single Man, Though in a Polygenetic Situation*

The answer to the first question must be that it is possible, even in an originally polygenetic mankind, that *one* man decided the issue whether human descendence coincided with a communication of grace or not. But first of all we must again eliminate that misunderstanding which treats original sin as if the specifically and strictly subjective guilt element were passed on to those that "inherit" because they are *physically* descended from the one who committed this subjective sin. Since *this* subjective guilt can in no case be passed on (whether we think of man's origin as monogenetic or polygenetic), the question can only be whether the personal guilt of one individual within the original group of human beings can be thought of as blocking the grace-transmitting function which accompanied human descent from this group.

The answer is simply, why not? If man is inextricably both personal *and* communal and both these aspects *presuppose* each other, if the original group (monogenetic or polygenetic) is in any case a biological and historical unit and this group in any case determined in many ways as a unit the further development of mankind in its existential situation, then there is no reason why the decision of one individual *within* this unit could not influence the grace-communicating function of this group. If one asks how we can know this without just imagining it, we may simply answer that we know that we receive grace *only* through Christ as source and as the one who transmits and, on the other hand, that such a transmission had to exist from the origin of mankind.

One can insist on this second point. But we know this from revelation which clearly implies that all grace comes from Christ. Moreover, if all grace is grace of *Christ,* one can still maintain that a grace-transmitting function accompanies human descent, particularly with reference to two points contained in the New Testament. First, there is the overall solidarity of mankind in redemption and salvation history. This implies that one receives the offer of salvation only insofar as one is a member of this mankind. Secondly, Christ's saving function is always seen in scripture as concerned with man as *sinner* and not with man as such to whom grace was not due as something supernatural. If scripture sees this sinfulness as something personal, one cannot generalize this aspect—think of children, for example—and therefore we must accept a certain sinfulness which precedes the personal decision of the individual.

To this one may add that the historical situation of salvation is always influenced by the decisions of individuals. To understand this there is no need to hark back to those horrifying juridical contrivances which beset so many theories about original sin since the 16th century, according to which "Adam" was given a mandate for mankind (in our context: the original group) by a "decree" of God. This strange notion rested, at least by implication, on the false premise that Adam's subjective guilt passed to his descendants and, above all, isolated and absolutized the individual so that it then had again to fill the gap between these individuals by more divine decrees.

If, however, we see the individual, not only biologically but also historically, and in spite of personality and freedom, precisely as the member of one mankind, so that any individual decision *already affects* the situation of all men historically, then we can well dispense with this kind of "decree" theology. All we need is the premise that only a *whole* sinless original group can transmit grace to its descendants. That things can at least be conceived as possibly having happened this way seems clear. That the situation is actually like this, we can only know for certain if we:

(a) presuppose a general sinful situation of mankind which precedes the personal decision of the individual, as part of revelation;

(b) reckon with possibility of a polygenetic origin of mankind; and

(c) want to avoid the assumption that every individual of the original group was personally a sinner.

I want to emphasize two points for further clarification. *First,* personal sinlessness as such is no claim to a grace-transmitting descent from such a sinless person. Innocent members of the original group can (or must) be thought of as grace-transmitting only insofar as they are members of this one mankind (of this original covenant of God with mankind). If therefore this mankind is influenced by sin (and this is clearly thinkable even if only one individual sins), then this original group ceases to be a transmitter of grace, and this holds also for all individuals, regardless of whether they are personally innocent. *Second,* even a polygenetic mankind, when conceived as a biological and historical unit, must not be understood as a homogeneous mass of equal individuals. In every aspect of human existence this unit may contain individuals who determine in a special way the common historical situation of all and so "represent" this unit and situation. This makes it still more possible that even one individual can block the grace-transmitting function of the original group, even if we would not accept that every sin of any member of this group would have this effect.

6. *"Adam" as the Collective Expression of the General Guilt of the Original Group*

In answer to the second question, we may say that it is quite thinkable that the original group denied God in all its members at the beginning and that all together thus compose this "Adam" who blocked the grace-transmitting function of the original group for its descendants. If we assume in this way a *collective* sin in the original group, then there appears to be no real difficulty at all about a polygenetic origin insofar as the teaching on original sin is concerned. "Adam" as sinner then is the concrete expression used for that one group, which originated mankind, also in terms of salvation history; the group has sinned as a whole, with the consequences which traditional teaching attaches to this sin.

This view in no way implies any Pelagian or Erasmian interpretation of original sin. The sin of original mankind does not spread by imitation but via generation (DS 1513f.). This original group is clearly contrasted with following generations and, insofar as the grace-transmitting function which ought to accompany descent has been blocked, this group creates an unredeemed situation, which is rightly called "original sin" and which precedes the personal decision of those that come afterward. Moreover, the assumption that all men belonging to this original group were sinners is not as arbitrary as appears at first sight. Even if we maintain that the group and not the individual (capable of procreation) is the basic genetic unit, the polygenetic theory admits that this original group must have been very small, particularly if we exclude those that never reached the age of historical and personal decision. Even present-day evolutionary monogenism accepts a "group" (namely, two) that sinned. When we remember that Catholic teaching holds that in the state of original sin mankind (and this in huge numbers) could combine *general* sinfulness of a personal nature (at least in the matter of sins of omission) with *freedom,* then such a view is certainly not arbitrary where the small number of the original group is concerned.

The difference in the matter of freedom before and after the sin is not a decisive element here. Otherwise it would be impossible to think of sin at all in man's existence without this sin. In agreement with what has been said above, the freedom of the original group must here, too, not be seen as the freedom of isolated individuals who each turn from an individual, absolutely neutral and even positive freedom all together and suddenly to evil. Here, too, one must think of a freedom situation in which all share and one influences the other in his individual decision, without forcing him.

V

CONCLUSION

It would at least appear to be neither certain nor necessary to maintain that only a monogenetic original group (one individual or one couple) must be at the origin of that first sin of mankind in order to explain what we call original sin in the orthodox and traditional sense of the word. A polygenetic view of mankind's origin also allows for that first sin, whether individual or collective, which created that unredeemed situation for the whole of mankind that came after them, and which we call original sin. But it seems to me to require that this original group be a biological and historical unit which is also significant for salvation history. This requirement seems to me also fulfilled in terms of polygenism. Therefore, there seems to me no reason for the magisterium to intervene in the matter of polygenism in order to protect the dogma of original sin.

Heimo Dolch/*Bad Honnef, W. Germany*

Sin in an Evolutive World

One of the most frequent and perhaps most important objections raised against the idea of a totally evolutive world, such as has been presented in Catholic thought in recent times by Teilhard de Chardin among others, is that the idea of sin would lose serious meaning. It would no longer be a freely willed denial of a law recognized as divine, but a "waste product" within the one evolutive process. Teilhard may claim to have found an "intelligible and plausible" solution to the problem by seeing evil as the "statistical need for disorder within a given complex regarded from the point of view of its organization",[1] but this solution must be rejected because the dogmatic conception of sin would thereby be more or less completely undermined.

The following observations attempt to examine this question, which may be approached in two ways. Teilhard's statements may be examined in their totality and an attempt made to validate his remarks on the specific problem of sin.[2] Rather than do this I have opted for a more general approach and shall ask: (a) What is the basic structure of an evolutive world (or world view)? (b) What is man's place in such a world? (c) How are

[1] Teilhard de Chardin, "La pensée du Père Teilhard de Chardin," in *Les Etudes philosophiques* 10 (1955), p. 581.

[2] Cf. Ch. 4, "Das Kreuz und das Ubel," in G. Crespy's *Das theologische Denken Teilhard de Chardin* (Stuttgart, 1963), pp. 149-84.

his actions (and his wrong actions) to be characterized in relation to these?

Since the questioning will be of such a general nature, I shall not adopt any specific attitude to what Teilhard says, but merely refer to him for purposes of clarification. That it is at all possible to pose these questions in so general a way is due to the fact that for the majority of Christians today (whether they have an exact knowledge of the scientific theory of evolution or not and whether they agree with Teilhard's work or not) the act of faith no longer depends on their being able to reconcile belief in creation with the theory of evolution, as if two independently conceived structures, historically shown to be in competition with each other, had somehow to be synthesized. This is a situation of the past.[3] The theory of evolution is the more or less generally accepted intellectual background against which it is asked whether, and in what way, belief in creation can be given a content.[4] (Whether this is as it should be is not our concern. This is the situation we have.) And there is no doubt that one of the most disturbing questions which arises is whether, in a world of process (be it scientifically determined or historically conditioned) in which the whole reality of man becomes existentially questioned or analyzed away by depth psychology, sin does not simply disappear.

I

THE BASIC STRUCTURE OF AN EVOLUTIVE WORLD

How are we to describe the basic structure of an evolutive world in brief outline, since the space at our disposal precludes doing more than this?

[3] H. Volk's *Rektoratsrede* (Münster, 1955). "Schöfungsglaube und Entwicklungstheorie" was undoubtedly an important milestone on the path of dogmatic reflection, just like the *Rektoratsrede* of F. X. Kiefl (Würzburg, 1909) half a century earlier. The point of milestones, however, is that they indicate the way ahead.

[4] We are taking the idea of "belief in the creation" as an abbreviation for everything that the Christian believes about God, himself and the world and the fulfillment of the latter two in him and through him.

Perhaps one should first answer negatively. If we try to define the particular quality of this world view by saying that the element of change or growth in general and of evolution in particular must be more strongly emphasized as against the static world view of antiquity or the Middle Ages, we are missing the central element of this new conception. It is not a matter of change of emphasis, but of a fundamental change of attitude. F. Dessauer has brought out this difference very clearly. We shall quote just one passage of his and leave it to the reader to go further into the subject if he wishes: "In the new view of things that came with Galileo . . . within the structure of perception and of knowledge, the physical, changeable object naturally remains the most important thing in physics; but in the dependent structure of being, the causal nexus process, it no longer enjoys this primacy. Objects are changeable products of the operation of 'forces', which thus precede them and— this is the important thing—wherever they appear they operate according to laws that are immutably valid throughout all experienced time and space. These laws of cause and effect are what is primary." [5]

In this view of things, then, it is the laws or principles that come first, in every sense; in second and subsidiary place come the objects, the physical bodies. The former span the realm of reality in which the latter are understood as concretions, agglomerations or constellations. E. Nickel describes this basic change of attitude in the following way: "According to our conception of the world, reality is to be described in all its breadth and depth as the actualization of all 'previously conceived' structures. It is not materially constituted elements from which all else is derived, but elements of 'pre-material' being, which can be actualized in matter or in other categories, according to the way they are 'determined'." [6] These previously conceived

[5] F. Dessauer, *Der Fall Galilei und wir* (Lucerne, 1943), pp. 44-6. Cf. *idem, Naturwissenschaftliches Erkennen* (Frankfurt, 1958), pp. 216-23.
[6] E. Nickel, *Zugang zur Wirklichkeit* (Fribourg, 1963), p. 130.

structures are the determinants of what comes into being,[7] the "power lines of reality".[8]

We recognize that, precisely because in an evolutive world these principles, being-determinants and power lines are primary, it is not enough to add the dynamic element to the world of objects to give it greater importance. No, we must start with this idea of the dynamic. The world is process; it is a stream of being.

Insofar as we recognize these principles, we know how the world is process, but not its origin, its goal or its course. To ask this is to ask a theological question, whether as a scientist or as a theologian. The two possible answers, both totally different, are immediately clear. Either one believes, with Einstein[9] or B. Rensch,[10] in some such pantheism as Spinoza's, or else one believes in a theistic, revealed faith. According to the latter the world came into being *ex ordine sapientis* that it might "represent" in diversity and variety what had always existed singly and undivided in the divine goodness.[11] And if St. Thomas sees things (and their natures) as manifestations of the one divine plan, then one may do the same in relation to the principles or being-determinants.[12] In and through them is performed what God wants to happen.[13] Because he wills and inasmuch as he wills in and through these determinants of being, he is

[7] *Ibid.*, p. 129.

[8] N. Luyten, *Teilhard de Chardin, Eine neue Wissenschaft?* (Freiburg/ Munich, 1966), p. 54.

[9] Cf., for example, A. Einstein, *Mein Weltbild* (Amsterdam, 1935), p. 21.

[10] Cf., for example, B. Rensch, "Die philosophische Bedeutung der Evolutionsgesetze," in *Die Philosophie und die Frage nach dem Fort-schritt,* ed. H. Kuhn and F. Wiedmann (Munich, 1964), pp. 179-206.

[11] St. Thomas, *De pot.* III, 16 c.

[12] Phrased more exactly, these determinants of being are partial manifestations of the one divine plan, since according to St. Thomas this copy is not unambiguous. *"Omnia quae praeexistunt in causis realibus, praeexistunt etiam in divina praescientia, sed non e converso, cum quorumdam futurorum rationes Deus in se retinuerit, rebus creatis eas non infundendo . . . .": De ver.* XII, 3 c; cf. *De pot.* III, 16 c: ". . . non tamen omnis finis est forma."

[13] St. Thomas, *De pot.* III, 7 ad 16.

always active.[14] Thus the world is not something that just exists and changes under certain conditions; rather, it is essentially "on the way": "I am the way, the truth and the life" (Jn. 14, 6).

Perhaps we have not sufficiently pondered these words. Jesus is not only truth for believers, but truth in itself; he is not only life for him who believes in him, but life in itself. One could surely argue in similar fashion that Jesus is not just the way for all who seek in him the way to the Father, but is in his very nature a way. And does that not make everything that was "created through him and for him" (Col. 1, 16) like a way?

II

THE PLACE OF MAN

If we now seek to answer the question of what the place of man is in an evolutive world, we shall no longer be able to see this like Descartes as a simple single point, as the *res cogitans* as against the *res extensae*. Man also is, just as much by chance as by plan, a concretion; the motivating forces of the evolutive world pour through him also, shaping him. The results of research in biology, the study of behavior, anthropology, etc., have compelled the recognition of these facts, but they have also showed something that is sometimes given scant attention by theologians when discussing the subject: namely, that the evolution of species does not mean the removal of their differences, but the attempt to understand their origin.

According to the general view (cf. the statements of J. Huxley, G. Heberer, B. Rensch, etc.), man does in fact march at the head of evolution (the question whether this was an orthogenetically intended goal need not be answered here)—i.e., man is always a "particular formation" (G. Heberer) of the fundamental entity, "life in matter and realizing itself in it", but not just any old formation. He is unique, since in him—and, as far as we know, only in him—evolution comes, as it

14 *Ibid.*

were, to itself; he becomes aware of himself and of it, and re-
flects about himself and it. The great principles of composition
and decomposition, of order and disorder are also active in him,
but in a special way: not as entropy and ectropy as on the level
of inorganic and organic being, but in a human way. Consciousness (the principle of order, according to Teilhard) rises to the
level of reflection, and the general drive toward realization rises
to the level of conscious formation.

This is the second point that has to be more closely examined.
Even the most extreme theoreticians of evolution do not deny
that there are such things as consciousness, decision, responsibility, etc.; they only think that they are able to give sufficient
reasons for them with one particular method and set of laws.
Against this both the philosopher and the theologian rightly
argue that man is not an object that can be adequately investigated by scientific means alone. But this should not lead one to
assume that the existence and the nature of consciousness
(mind) were always denied by scientists.

They were certainly not denied in the view of Teilhard, as
perhaps a few quotations will make clear: "We are all faced
today with a problem that is, in my opinion, more important than
any technical question of how to give shape to the world, namely
that of value, inasmuch as we desire to face with full consciousness our destiny as living beings, i.e., our responsibility toward
'evolution'. Further down the river that is carrying us, a whirlpool is starting to build up. . . . This whirlpool is undoubtedly
stronger than we are. Yet in our quality as human beings we are
able to judge it in order to steer our way through it. I should
like . . . to examine the various directions that are open in
this critical moment to him who is holding the tiller, that is,
every one of us. And the final decision as to which is the best
way will be the *big decision.*" [15] Every new piece of knowledge
shows us new determinisms and imposes on us a further "burden

[15] T. de Chardin, *Die grosse Entscheidung, Die Zukunft des Menschen*
(Olten/Freiburg, 1963), p. 61. As we have taken the subsequent quotations from this collected volume, we have mentioned the particular page
numbers in the text.

of continuing the world", which undoubtedly weighs "ever more heavily on the shoulders of humanity" (p. 62). "Since we have become men, we have acquired the capacity to plan a future and judge the value of things; [therefore] we can no longer act without our very refusal to take up an attitude being the equivalent of a decision" (p. 69). We men, who have crossed the "equator of hominization", have to bear the burden of this knowledge. Human unity can only last and grow through an "acceptance from within", not from a "compulsion from without" (p. 249). Thus Teilhard demands an "increase in passionate concern" (p. 158) for the shaping of the future; he calls us again and again to make "the great decision" (pp. 57-85) and asks what strength we can ultimately "gain from the *pleasure* of advancing despite the shadows of death" (p. 309). Are not "these forms of attraction that posit the possibility of our eventually becoming one ultimately connected with the radiation from a final center (both transcendent and immanent) of psychic energy—and of that center whose existence, since it opens a door for human activity into the irreversible, seems indispensable (the supreme condition of the future)"? (p. 309).

III

THE RESPONSIBILITY OF MAN

Thus—in a totally evolutionary world consciousness—intellect, the possibility of decision and responsibility are by no means denied by an inner necessity. One may not totally agree with Teilhard in his statement that "a world of convergence . . . whatever sacrifices of our freedom it may seem to demand, is the *only* one" (Teilhard's italics) that "saves the dignity and the hope of being" (p. 75)—still less if he concludes from this alone "that it *must* be true" (*ibid.*). Nevertheless, one may still say that, in view of the mentality of contemporary man, this world view—and this one supremely—is able to make clear to him his dignity and task. After all, he asks to be able to support his views as far as possible by experience or the results of scientific research.

Certainly, man is no longer to be naively considered as the central point of a statically conceived world (here it still remains to be investigated whether the ancients really saw their world as statically as Teilhard and many theoreticians of evolution maintain), but that does not mean that he has lost his kingdom. The very insights of science in general, and those of evolution in particular, teach him in detail to see this kingdom of his tangibly in the "phenomenon". He does not possess it, but he wins it by fulfilling his task responsibly.

Teilhard wrote very early: "The true call of the cosmos is an invitation to take part consciously in the great work that is going on in it: not by descending into the stream of things do we unite ourselves with their unique soul, but by striving with them for a goal that is to come" (p. 13). A little later he stated his belief that "our true kingdom consists in serving, as intelligent atoms, the work that has begun in the universe" (p. 30). If the world is a stream and flows not from some obscure origin somewhere to some vague goal, but is a creaturely emergence because of a divine immergence, and if, further, man in it has consciousness, then his task is not to be driven along in the stream by natural laws, nor even to affirm being placed in the great stream in a Spinozistic *amor fati*, but to decide, to take up an attitude.

It is true that here there are fundamentally only two choices, two attitudes possible, joining in or not joining in. In the former and through it he fulfills his task and also draws nearer to his fulfillment; in the latter he fails in it and refuses himself. But because man is a concretion in the specific here and now and his freedom is thus always "situated",[16] the two basic attitudes are realized in specific, different realizations of objects, in products. If they are in the one order, they are what is required; they are good. If they fall short of this order, this requirement,

[16] Cf. P. Schoonenberg, "Mysterium iniquitatis," in *Wort und Wahrheit* 21 (1966), pp. 577-91, esp. p. 580.

then they are poor or waste products. We are "as little free in the conduct of our lives to follow blindly our inclinations as a ship's captain could follow his moods to decide the way to the harbor" (p. 70).

IV

Sin in an Evolutive World

We see then that in an evolutive world sin does not lose its character; that depends on the particular understanding of this world. At most, one might say that sins are ordered, as it were, "hierarchically". There are not a large number of sins strung out horizontally, but only one real sin and its various observable effects. The real sin is the refusal to cooperate, the denial, realized in the action of the basic nature of man: to be the captain of a ship who has to bring it (and hence himself) through its voyage, in the strength of God, to the saving harbor of the *parousia*.

I am aware that with this not all the questions raised by our subjects are answered. I only wanted to make a few basic observations.

Whoever maintains that sin loses its serious character in an evolutive world view, makes in fact the same mistake as those who deduce from the existence of suicide statistics the falseness of the thesis of free will. Because we can predict with statistical probability how many men will commit suicide on a particular day, there can be, so it is argued, no freedom in this respect. But this argument fails to see that these statistics—unlike, for example, the statistics for road deaths—only tell us about those deaths that are voluntary, and not about those that occur through accident or illness. These statistics, then, do not eliminate freedom, but wholly assume it! Similarly, we can certainly speak of sin in an evolutive world view as of a waste product that occurs according to statistical necessity, inasmuch as it is seen as the phenomenal appearance, the effect of the voluntary refusal.

Dominique Dubarle, O.P./*Paris, France*

Does Man's Manner of Determining His Own Destiny Constitute a Threat to His Humanity?

This question is extremely vague and can be discussed only by bringing into play a host of concrete precisions. These will then allow a partial reply of some kind. But it is extremely difficult to derive from them a general reply possessing any real philosophical or theological value. In any case, before dealing with the matter itself a few rather elementary facts must be recalled.

I

THE ESSENCE OF MAN THREATENED BY MAN

The Essence of Man—Result of the Historical Process

Man is "by nature" an animal arising from an animal nature. He is by the same token inevitably called upon to transform himself ever more for the better through the very effect of his specifically human development. This transformation is an historical process involving at least three factors: natural animal conditioning, specifically human energy and lastly, the concrete immensity of circumstances—those of the natural milieu as well as those of the human environment itself. Whatever may be the case with these factors, man is himself only when pursuing, so to speak, his own essence, progressively attaining it starting from

85

his initial human nature, and making the most of his concrete conditions and circumstances to bring forth results.

Hence, far from having been given in completed fashion at the outset, the essence of man can only be the result of an historical process. Man then does not know in advance what this essence will be. At least, he does not know it in an absolute sense. At the end of the process, whatever has taken on the value of the essence will be partly the result of his own will but also partly the product of the external predeterminations and historical risks with which his will is concerned at every moment.

For its part, man's specifically human energy can intervene at least in two different ways as a component of the historical process. It can appear as an already intelligent and voluntary —hence human—energy; but one that is still primarily spontaneous, incompletely conscious of self in its intellectual developments and practical involvements. As a whole, such is the energy that collectively sweeps along that part of the present world which lives off scientific research and ultimately off the social realization of science.

However, it can also appear as an energy that has reached a higher degree of lucidity and self-control and one that is alone really free to choose its ways, instead of obeying its spontaneous impulses without truly mastering them. Such is the case in relatively well circumscribed instances wherein it is permissible for the individual or the group to establish for itself a distinct series of possibilities; to deliberately choose a particular one from these rather than some other, and then to see where the choice leads and in a pinch to modulate pursuit of the action, upon sight of the consequences progressively manifested by the initial choice.

Man Attains Himself Only at the Risk of Losing Himself

Thus, the act by which man wills himself has as its primary and principal problem the passage from what is done out of a still spontaneous impulse to what might be done at the level of the fully deliberate choice of self, capable of wielding control

over subsequent action. For the delivery of self over to spontaneous impulses of the will, to passions lacking the flexibility
of their élans, does not always and everywhere lead to happy results. But even supposing that such a passage has been achieved,
a second problem arises. This is the problem of attaining the
desire for what is truly good and capable of happily constituting
the being who will represent the conclusion and ratification of
the entire process.

At a certain point, at the very least, such a problem is in no
way simple. For man does not know perfectly in advance what
will constitute the well-being of the just man in the concrete;
hence, he does not know perfectly in what and how to will himself. He can desire to escape certain evils which life has made
him experience, take account of his previous conditions, and
have regard for circumstances; but none of this is yet objectively
and absolutely determinative for a freedom. To know in an
ideal and detached way what constitutes this well-being does
not otherwise prevent the will from affixing its choice in a
manner that is extremely inconsonant with this knowledge—
either out of the very human need for less difficult behavior or
even under the influence of what is termed the desire to do evil.

For all these reasons, which concern the nature and the procedures of human energy, humanity—the properly developed
essential being of man just as much as the dominant existential
satisfaction of the collectivity—is in constant danger from man
and his willful energy. Historical man cannot exist or act or treat
himself in any fashion whatever without running at least the
partial risk of failing and ultimately destroying himself. A few
concrete applications of these truisms will enable our thought to
concentrate on actual cases which we discuss with some anxiety
among ourselves. These are furnished by the basic forms of the
human utilization of science.

II

The Classical Socialization of Actions Animated by Science and the Present Attraction of Its Expansion

In uninterrupted succession, what was set in motion two centuries ago (period of the first industrial use of the steamboat) has brought us today—at the end of an extremely vast development—to the two planetary problems embracing the situation created by nuclear arms and the economico-demographic evolution all over the globe. The human energy responsible for the emergence of these problems and the tendencies that will be increased rather than resolved is a collective energy of a "spontaneous" nature—at least, so long as our evaluations involve the human genus as a whole.

This energy continues to advance the process (which seems to be accelerating) while still incapable of grasping itself lucidly and of formulating a coherent choice on the level of the human world. Still less is it possible for it to control the consequences of its choices. The fact that it is of such a nature and that it will continue to be such on account of historical inertiae, constitutes an ever-increasing danger for the present civilization and the whole of humanity which it claims to be advancing to a better state. It is never stated that the human process, as presently developing, does not risk—upon its more or less proximate "coming of age"—the breakdown of its advancement, and the confusion and dispersion of all that it seems to have permitted to be attained.

Failure to understand this very positive danger could scarcely offer proof of wisdom. What the *Constitution on the Church in the Modern World* has to say in regard to the major problems of war and poverty at the national level constitutes a solemn warning on the part of the Vatican Council II.

"The unique hazard of modern warfare consists in this: it provides those who possess modern scientific weapons with a kind of occasion for perpetrating just such abominations; more-

over, through a certain inexorable chain of events, it can cata-
pult men into the most atrocious decisions." [1]

"Warned by the calamities which the human race has made
possible, let us make use of the interlude granted us from above
and for which we are thankful, to become more conscious of
our own responsibility and to find means for resolving our disputes in a manner more worthy of man." [2]

"Humanity, which already is in the middle of a grave crisis,
even though it is endowed with remarkable knowledge, will perhaps be brought to that dismal hour in which it will experience
no peace other than the dreadful peace of death." [3]

From the present point of view, we must, however, add the
following remarks:

The Possible Danger and the Consciousness of this Danger

1. The danger presently in question concerns chiefly the
economico-socio-political system. This is a very real component
of the principal goods that humanity must gather together in
itself. But as such the component is not yet identifiable with
what is at the very basis of all truly human mankind. Defeat on
this level in no way necessarily constitutes a decisive disaster
for man. Great civilizations can collapse and their principal
goods become relegated to disorder and the scrap heap. Life
continues nonetheless.

Given time, later generations are able to bring forth something new from the riches of much of the ruins and a few new
inspirations. The years ultimately wipe out catastrophes; and
considered in this light, such catastrophes can even present some
aspect of necessary collapse and salutary upheaval.

Accordingly, if our civilization should have to pass through
some analogous collapse, we would still have to trust in the
animal resource of the human race, the relative clemency of the

[1] *Constitution on the Church in the Modern World,* n. 80.
[2] *Ibid.,* n. 81.
[3] *Ibid.,* n. 82.

universe and even the collective will—profoundly injured perhaps, but still capable of healing itself and drawing some instruction from the experience thus undergone.[4] Even more, we would have to trust in God.

2. At present, we must acknowledge that it is hardly natural for man to suspend the spontaneous impetuosity of his impulses so long as their direct result does not appear to be an insupportable evil to him. Everything continues in the same vein until the moment when the ultimate of evils and the abyss are within sight. That is when man calls a halt. Reality and history are so constituted that on the whole man has, up to the present, stopped short of definitely compromising his existence and the faculties for the resumption of his progress. We can therefore venture the belief that the same will also be true in the future. Moreover, if the curbing and reconsideration of its present impulses are in no way the reaction of humanity in full scientific ferment, the reason is most likely because its collective conscience does not yet flash the imperative warning of intolerably painful malefactions and terminal dooms.

3. However, we must temper somewhat the assurance that we might draw from these latest considerations. First of all, one of the major present ills of mankind is precisely the fact that scientific development is as a whole something euphoric only for those actively responsible for it, while its foremost bad effect falls primarily upon peoples that have remained more passive than active. Consequently, the part of humanity most responsible for the present deterioration of the whole still perceives only dimly —in an external and detached way—what should already alarm it and would certainly do so if it were itself experiencing it directly.

In the second place, we must also take account of the uncharted nature of scientific energy. In principle, science is the faculty of methodically efficacious action to infinity in its proper

[4] *In the present state of things,* we must exclude as completely chimerical the human possibility of totally suppressing every human species from the face of the earth. Even by utilizing the total amount of nuclear explosives presently available this could not be achieved.

line of efficacy. At the moment when it becomes animator of human evolution, it suppresses man's protective element against himself, formerly constituted by the ensemble of natural limitations on prescientific technology. Once this protective element has disappeared, it may well be that man might imprudently maintain or heedlessly base his action upon courses that science will render fatal for him, but which would not have been such if the available means had remained those anterior to science.

Certainly, it is not a question of saying *a priori* that this is what will happen. It is even a good deal less fatal that science in man's hands might compromise human development. Yet it remains true that this calls forth on the part of humanity a collective virtue which was not needed yesterday and whose attainment cannot easily be forecast.

III

SCIENTIFIC ACHIEVEMENTS RELATIVE TO BIOLOGY AND HUMAN ACTS

The domains of biology and the human sciences henceforth offer many opportunities for advocating an increase of collective prudence. For it is at this very moment that, strengthened as it is by a rather considerable achievement, science is on the way to taking giant steps forward as regards knowledge and probably, as a consequence, the capacity to act. But in this case the malefactions would await man from within himself, diminishing him physically, altering him psychically and falsifying him culturally and ethically.

Progress of Biological Sciences

We will rapidly summarize the achievement in question. For almost the last two centuries the biological sciences have carried out this first exploration of facts which has given man on the one hand the possibility for effective medicine, surgery, and hygiene, and on the other hand an increasing mastery in the domestication of the animal and vegetable world, to the benefit principally of food production. The progress of agricultural technology and, even more, the present demographic expansion are the firstfruits of this achievement.

For their part, the human sciences have enriched and refined the psycho-physiological understanding of mental life—in particular, its formation in childhood, social motivation and pathological modalities. They have also given solid footing to valid use of mathematics concerning collective attitudes and behavior beginning with the field of economics that in our day already exhibits a notable capacity for controlling processes.

The result of all this is threefold, at least in a milieu of advanced civilization. An indisputable better life biologically speaking—nourishment, health, etc.—has been achieved. The disciplined conditioning imposed by society on the individual under the heading of things necessary for the good of the whole has been made sensibly more explicit. Finally, numerous cultural resources have been put into circulation even at the level of the mass of humanity; these are facts collectively available even though the absence of sufficient leisure time and the required education does not allow everyone to derive all that is desired therefrom.

It seems apparent that the biological sciences are going to succeed in deciphering the principal constitutive modalities of life beginning with elementary terrestrial materialities. It is not impossible that within a short time the ensemble of biological laboratories will be in possession of all the material links whose integration expands the reality of the living organism. Elementary productions of living forms could then be envisaged, variants sought from the evolutions experienced.

In like fashion, modes of acting on already existing organisms—including that of man—can be entirely renewed. What has for some time already been achieved with the possibility for chemically controlling the natural processes of feminine ovulation provides only a first and perhaps very faint idea of future possibilities.

Progress of the Psychical Sciences

On their part, the sciences of the psyche and human actions can also experience spectacular progress. The basic conditioning—both physico-chemical as well as neural and cerebral—of mental life is explored a little more each day. Pharmacology, neurology and psychology see their knowledge expanded and—in an admittedly still limited manner—their powers of acting as well. In correlative fashion, certain collective inhibitions give way or rather become transformed in the vortex of modern societies which for better or worse no longer have the same requirements in mores as yesterday: for example, relaxation of harsh punitive measures and liberation of eroticism.

On the other hand, scientific and social statistics have brought into play a whole set of new instruments, from giant computers to the mass media of communication: cinema, radio, television, etc. Nothing about all this yet forecasts absolutely decisive conceptual changes; if these are one day to emerge, they are probably much further away from us than the great changes in biology that would achieve the scientific synthesis of life.

Whether these be near or distant insofar as their completely decisive character is concerned, such achievements are characterized by the possibility held out to man of scientifically acting upon himself in a direct fashion. Man can modify his body, the natural equilibriums of his health and illnesses and the conditions of his reproduction. Soon he may be able to influence his ontogenetic development. Man can act on his psyche both by ingestion of various substances whose list grows longer as well as by multiple intentional conditionings—from individual temperament or education to collective indoctrinations and attractions.

The scientific control of social life—something that has now become indispensable—has very profound mental and cultural modifications as its results. In other words, modern man is prey to an excessively multiform treatment of himself by himself, a treatment that is constantly haunted by the proposal to increase his efficacity through science and in which the best and the worst elements may very well be found—as a result of individual initiatives or collective enthusiasms.

Optimism or Pessimism?

We can thus envisage a humanity rendered better by the manner in which it transforms itself under the direction of science, along the general lines of the scientific optimism of the 19th century, heir to the philosophy of the Enlightenment; or we can envision a humanity which, by wishing to put into practice the resources of science in an inconsiderate way, would fall victim to the plague of a radical sophistication of self. This is the idea that rules a number of modern novels of the future, for example, the *Brave New World* of Aldous Huxley.

The idea that such an eventuality is possible has its present usefulness, for the Christian as well as for the philosopher. It fittingly completes the flattering tableau of the possibilities of science by more somber images. Thus we see renewed, and placed at our disposal, the ancient Platonic method of two opposing models for realizing existence: the one happy and the other totally permeated with extreme misery. Yet, nothing would be worse than to make a panicky reflection from such images of human defeats, to hypnotize oneself for today by the symptoms judged unfavorable and to paralyze action by the false conviction of an irrevocable march toward evil.

For this type of pessimistic dramatization we should first of all recall that at the moment when it becomes impossible for man to cling to past attitudes, he does not really yet know completely what his humanity should and can be. The only recourse for each age is to assay what it might well be, released from reerecting the attempts made in any direction the moment it becomes really evident that they do not and cannot lead to any good. So is it and so will it continue to be regarding the use of the new knowledge that the biological and human sciences yield for us.

Doubtless, the human modality of the march toward the future is that of an action pursued in the name of concrete aims, in

which good and evil aims never cease intermingling and dis-
placing each other in alternations of causality; as a result, the
consequences of actions are paradoxically contrary to objectives
sought. The effort of goodwill leads to such or such bad effects;
evil behavior gives rise to such or such a favorable result. When
this takes place on historical dimensions, it does not engender a
happy state for man. Yet it is a long way from the needy and painful history of man—which is the permanent condition of the
terrestrial reality—to that cursed history about which an appre-
hensive pessimism writes, even though many aspects of human
affairs readily suggest models for us.

The Christianity of a Humanity in Search of Its Essence

Finally, and this is not the least proposal in the present consideration, we must needs confront the Christian with his own
faith on such an occasion, inviting him to refine its teachings.
There is evil in human action and in one way or another this
inevitably harms the very humanity of man at the moment when
human works yield results to their authors, and more and more
man makes himself the object of his own actions. The injury to
man risks becoming deeper or more pernicious at the moment
when the action of man disposes of the resource that we call
science. And yet this evil and the wounds it causes are conjured
up in their principle by Jesus Christ and by all that becomes one
with him in the midst of the human world.

This shows us the true role of the Christian here below, salt
of the earth on condition that he does not lose his flavor, and
charged with achieving in each generation what still remains to
be accomplished in the redemptive action of Christ. However,
we must not be surprised if the success of the Christian enterprise is never more than partial and highly imperfect in the
context of our history. Christian success can only materialize for
a time. It must afterward be renounced and dispersed anew, as
has happened in what we imagine to have been the humanistic
success of Christianity.

What is more, throughout the length of the millennia, the redemption of humanity remains equally contemporary with what St. Paul calls the "mystery of iniquity". The historical progress of man in search of his essence, to the contrary, points in the direction of an eternal humanity happily fulfilled, and to what is the very negation of this. In this regard, humanity, yesterday as today, is prey to this risk of going against its humanity. In its ultimate form this is the risk of freedom, of the free will confronted even with the choice of accepting or rejecting grace, of consenting to the universe of the children of God or of searching, without God or against him, for some other humanity and a universe that is its reflection.

Jacques Ellul/*Bordeaux, France*

The Technological Revolution and Its Moral and Political Consequences

I

WHAT IS MEANT BY TECHNOLOGICAL REVOLUTION?

The first problem lies in the precise meaning of the word "revolution". Contemporary books, articles and conferences on problems of technology and of the technological world use the word incessantly, designating by it a wide variety of phenomena. It can mean, for example, an acceleration in the development of technological processes or of the opening up of the areas in which these could be applied; the word might also be used to indicate the basic fluidity of a particular situation that can lead as readily to the dissolution of structures as to their renewal; finally, it can mean an alteration in social and political structures.

Whereas these different situations might well be the symptoms of a revolution, they are not yet in themselves revolutions as such, for some phenomena can go through a calm and natural birth that does not need the prior impetus of a drive for freedom and justice to precipitate the process. There are other phenomena that have perfectly rational links with one another and are therefore to some extent foreseeable; they do not depend for their evolution on an invasion of history by irrational forces.

Furthermore I would say that as far as the phenomenon of technology is concerned, there exists neither that which Marxists call an objective revolutionary situation nor what the liberal mind refers to as a revolution.

The truth is that we have here a term whose meaning has been so distorted by passions and prejudices that it has been rendered unserviceable in a scientific discussion. For the purposes of this article it seems to me that one can at best speak of an accelerated evolution of technological progress.

II

THE POSSIBILITIES OF TECHNOLOGICAL ADVANCE

Let it be said immediately that such possibilities are unlimited. There is almost nothing that man cannot make with the help of technology. It is a process that knows of no foreseeable end to what is apparently an unlimited ability to construct every artifact of which man is able to conceive. Its nature knows of no check to its developmental potential. It is at the mercy only of external factors such as economic conditions, or the impossibility of producing the materials it needs, or the failure of scientific research to keep pace with technology's appetite. As technological growth requires the application of rational thought, the technologist is always able to see what lies around the corner. Because technological advance is a process of causality, not finality (the developmental impulse lies in the organized manipulation of elements whose availability is foreseen and not in the achievement of a particular goal), one is the better able to know and estimate what will be possible, or already in operation, in the year 2,000 or 2,100.

To state this is to note an important characteristic of technological development. The scientist or technologist does his job by using known and available elements, not by feeding exclusively on visions of the future. Whereas prophesying about some distant goal is a risky business, prophesying on the outcome of coordinating actual concrete factors is a good deal safer. There are available, therefore, a considerable number of reports from scientists describing for us probable technological developments that will take place within the next twenty, thirty or fifty years.[1]

In contrast, it is extremely difficult to predict the consequences of technological growth for political events and processes, for society as a whole, for the individual and, to a lesser extent, in the economic sphere, because technological growth alters all the traditional life circumstances and so makes its consequences more problematic. Further development in all these spheres depends upon technological advances in contrast to which any other impetus can be granted only subsidiary significance.

The difficulty of making predictions about the consequences of technology in the sphere of politics, economics and sociology seems to me to have two causes. First, it is because we must make two types of prediction. One will be about the technological development itself and the other will be about its effects. Given that technological growth unfolds itself causally, a prediction is made on the basis of a grouping of data, each individual item of which can be isolated and grasped with relative ease. But in the political, economic and social spheres, where a whole series of value judgments, hopes, plans and precedents have to be considered, no such situation obtains. Development in these fields is, therefore, never simply the result of introducing technology to a political or social situation, but the result of ideologies and technology acting upon one another. These developments, then, are never the result of an agglomeration of short-term advances, but one must rather, in general, discern a long-term development. The conclusion—that long-term predictions are more problematic than short-term ones—is obvious enough.

The other difficulty springs from what one might call the ambivalence of technological progress, which in my view is

[1] Cf., for example, the Rand Corporation's 1965 report and the report from the U.S.S.R. by Coutchev and Vassiliev on forecasting technological development (1964).

neither exclusively positive nor totally negative. Its absolute ambivalence consists not in the fact that its fruits can be turned to good or evil according to our judgment and discretion, but in the existence within itself of good and bad elements.

It would not be difficult to show that every technological advance, while it solves a few of our problems, also gives rise to others, usually different in type, which are most unlikely to be any more easy of solution than the first lot. It could also be shown that all technological progress brings with it some totally unforeseen and unintentional results that, after an interval has elapsed, burden us with further problems.

How does the Christian stand with regard to this problem of the scope of technological progress? One can safely say that there can be no such thing as a Christian influence on technology itself. Further development in technology is simply a part of technology's nature. Arguing about a fact such as that would be a waste of time. We should take good care not to start applying immaterial moral value judgments to technological growth. To say that it is good or bad would be as senseless as to say of a flowing river that its motion is good or bad. When all is said and done, we have just one factual situation to grasp: technology increases and it is up to us to steer it, alter it or frustrate it.

The first question we must ask ourselves, therefore, is how we pursue life in our new technological environment. The question needs subdividing. What are the probable transformations of social, political and economic life one can most readily predict through one's knowledge of technology? And the second part of the question is: Which of these changes can be to some extent controlled? Ready answers are not there for the asking.

III

PREDICTABLE AND DESIRABLE ECONOMIC AND SOCIAL CHANGES IN THE HUMAN ENVIRONMENT

Taking first things first: What is predictable? We are always tempted to search for long-term plans. For instance, we are tempted to make statements such as: When automation has reached its point of highest development, we shall possess more goods than we can use, shall no longer need to work, and so on. Such speculations were better avoided.

It seems to me that no truly scientific predictions can be made about what could happen when automation has developed to its fullest extent; it is simply impossible to estimate what alterations to life as we now know it might then occur.

In this respect there seem to be two lines we could pursue. The first of these is short-term prognostication. As Christians this concerns us directly. The questions we should ask are: What form will the transformation take? How will the change from a non-technological to a technological society complete itself? In what way will the change from a society lacking the fully extended resources and applications of automated processes to one possessing these to the full come about?

This, it seems to me, is a Christian concern, for here we are not speculating about an ideology or about preparations for distant eventualities, but about the life situation of contemporary man. It is a question, therefore, of asking what setbacks we might encounter. Automation breeds them readily and the immediate future holds numerous ones: appalling unemployment problems; industrial delays caused by disparity in delivery intervals (not all industrial processes or sectors of industry are equally open to the introduction of automation); the psychological problems of machine-minders predicted by many psycho-sociologists, wage calculations and settlements in a semi-automated economy.

These are but a handful of the weightier problems. Although it might be contended that I am citing isolated cases, the accusation would not be wholly just. There are 5,000,000 unemployed in the United States alone. These are the problems we ought to be solving rather than worrying ourselves unduly about what life will be like in 2,000 A.D.

Our second line of enquiry, having first evaluated short-term predictions, would be to consider how we wish the long-term future to look. What we can foresee is not necessarily what we want. In fact they usually stand in direct contrast to one another. In other words, when we see before us a desirable future, this is only because the future is in fact unforeseeable. For Christians to attempt to establish what is desirable, to build up a picture of what can or could be, is a proper and important activity. Nothing is more disappointing than the sight of Christians swallowing theories about society's future without themselves contributing anything specific to such theories. If as Christians we propose as the long-term ideal no more than democracy, or national independence, or a society that knows no poverty, we might just as well keep quiet, for we are contributing nothing.

To me it seems more important to proceed to a critical study of what is desirable in relation to what we can predict about the technological potential of the near future. Having formulated the desired developments, we would have to examine them critically to ensure that we were not striving after utopian dreams or goals that short-term technological developments would frustrate.

Notice that I do not say goals that technology pursues. Take democracy as an example of what I mean. It seems to me most important to ask if a fully developed technological society would not make democratic processes much more complicated than they already are, and that a fully extended technology would in any event imply a total transformation of the structures and forms which are normally said to constitute democracy.

In other words, we ought to proceed to critical analysis, and in this field Christians can play a decisive role in society, looking to the long term but ensuring that our vision is not out of touch with the predictable consequences of short-term technological developments. We shall then be in a position to contribute a revolutionary development to the developmental process of contemporary society.

From what has been said about the problems posed by a technological society, it should now be clear, provided that I have said it adequately, that long-term thinking is by and large necessary, that this must include a calculation of the consequences of technological developments in the short term, and that we must try to see how one proceeds from short-term predictions to a long-term concept.

You will no doubt be thinking by now that the problem is merely an abstract one and that the French still have nothing better to do than split hairs. However, I take the view that the situation in which we find ourselves calls first of all for the clear recognition of what is at stake and an exact understanding of it. Lack of clarity is unacceptable because in the circumstances it is inexcusable. At present we are obliged to compromise between what we know and what we want. When our world was but a world in which nature was rampant and over which we therefore exercised no control, where traditional values and procedures were hardly questioned, we adapted ourselves to it more or less spontaneously and instinctively. Formerly man was himself little more than a part of nature; his attempt to distance himself from it, to control it, led to his becoming truly man.

But nowadays no instinct or spontaneity helps us to live in our technological world. Far from being able to make any *spontaneous* adjustments, we have to do so through will power, knowledge and organization. And to the extent that we appreciate our obligations in this respect, we acquire the means to make the necessary adjustments. Necessity and potential go hand in hand. It is possible and necessary for our consciousness of the situation created by technology to increase. The object is not that the individual should feel that his own private knowledge, and therefore his control, is increasing, but that the group or society in which he lives should be aware of such increase. Generally speaking, society is now more ready to reason with itself, to reflect upon itself. Both the individual and the class to which he belongs have become more conscious of their situation, their limitations, hopes, needs and responsibilities.

Increased consciousness, however, does not automatically produce fruits that are necessarily applicable to our actual situation. We are quite capable of interpreting all the factors under discussion incorrectly. Consequently, a clear view of reality in every life situation is our primary objective, for without it we shall achieve nothing. Realism should be one of the characteristics of Christian thinking.

An increase in consciousness does not necessarily mean an increase in the ease and ability with which we adapt ourselves to our technological world. We are, and shall continue to be, perpetually aware of our lack of adaptability, but an increase in our awareness does presuppose a will to succeed in what we undertake. Decisiveness in this respect is vital and much more difficult to achieve than the following of instinct, for we shall be in the position of having to bring judgment and decision to bear in countless different situations.

We must also remember that every phenomenon in the field of instinct has now become to some extent a moral or ethical phenomenon. In the world of technology the apparent area of decision and the need for awareness has led to the emergence of an apparent ethical alternative for every question. And on these two planes, that of what is wanted and what ethically must be chosen, Christians would appear to have their proper function, provided they do not view the matter in question solely from the standpoint of the society in which they live.

IV

DIFFICULTIES AND DUTIES

At this stage of our investigation we encounter certain difficulties of an intellectual and ideological nature. I shall outline a few of them in brief.

First of all, we are inclined to reject traditional ethics and for that matter almost everything else history has handed down to us. It has frequently been maintained that what really hinders us from coming to terms with the technological advances of our age is our relationship with the past. There is a tendency to destroy everything connected with it. Such prejudices are extremely dangerous, for, faced with the precarious adventure modern life has become, and acknowledging the enormous store of information man is offered in a technological world, the attempt to abolish his intellectual, ideological and ethical norms could result in an ever more deleterious rupture of his personality, unless one simultaneously furnishes him with a new thought system, a new ideology and a new ethic that he is ready to accept, that permit him to survive and that enable him to coordinate all the knowledge that comes to him.

Everyone has minimum requirements as regards his view of life without which he will be unable to collect and collate the information available to him. He needs reality-related life concepts which, nevertheless, sufficiently differentiate themselves from reality to guarantee his distinctiveness in relation to technological realities, and at the same time maintain and perpetually renew the distinctiveness that differentiated him from the natural reality that was his previous environment.

I have already said that for the man whose life was little more than a part of wild nature, his environment was at once an intimate part of him and yet different from him. He had adapted himself to it and so attempted to alter his environment in order to adapt it to himself. This was what his brain was for. Faced with a new environment, it is our job to reenact the original adventure in precise detail. To do this man needs a well-anchored thought system (that science will not provide) which will enable him both to adapt himself to the world he finds and to preserve a certain distance from it.

In this respect it seems to me that it is the function of the Church and of Christian thought to strive on mankind's behalf not simply to facilitate the adaptation process, but to provide, as points of contact, starting points, the well-grounded general concepts so badly needed—and by concepts I do not mean antiquated reactionary opinions. The Church should give this matter further thought.

But this is another area in which considerable difficulty arises. This exists less in our obligation to change our ideas and concepts than it does in our duty to refashion our method of interpreting facts and events. Mankind, and that includes Christians, has traditionally judged from action, via its bearing, to intention. In reality, this method of assessing intention has been absolutely debased by the phenomenon of technology. It is simply inadequate to select an action and give it Christian motives.

If the basis of the foregoing analysis of the growth of consciousness—the recognition of choice and the problem of the relationship of the short to the long term—is correct, then judgments of events and actions according to their intention are no longer appropriate. The criterion should be their foreseeable consequences, and within the limits I mentioned one should make every effort to foresee them.

Some consequences will look dangerous, but these, too, should be clearly displayed. Others will appear to be as new as they are unacceptable, and yet others might conjure up visions of the extreme insecurity of our future. But the more dangerous and disquieting these consequences appear to be, the more they represent a challenge, in the sense that it is then up to man to master them. The larger the area of the unforeseen becomes, the more it is mankind's job to try mightily to reduce it. It is clear, then, that the extent of the unknown and the disquieting in what we can foresee is precisely the condition for true progress, insofar as such is at all possible in any particular case.

It would seem the one fundamental evil in contemporary social ethics is that by contending that all is well and will remain so, that technological progress is in itself a good thing, that technological development can be equated with ethical and spiritual growth, man is given a false sense of security. The lack of tension this type of euphoric state induces can be utterly debilitating, killing any likelihood that man will come to grips with his real situation.

If things really were as this view would have it, what reason would there be to look any further than the end of one's nose. Just give things their head and rest in the knowledge that "we've never had it so good". In fact, of course, to summon a man to assume full responsibility means to highlight the underlying ambiguity of the situation technology brings with it. Though I would certainly never wish to maintain that technology was to be deplored, I would insist that of itself it constitutes not progress but the opportunity for progress.

There is evidently a lot of thinking still to be done, a lot of research to get on with. This is not simply an area in which Christians can contribute, but one in which they are duty-bound to take the lead. This they will do with a thoroughness, an astuteness and a feeling for reality that human nature can command only through the guidance of the Spirit.

Emmanuel Mesthene/*Cambridge, Mass.*

Religious Values in the Age of Technology

I f we are to explore the implications of what is new about our age for the enterprise of theology and for the role of the Churches in society, it might be well to start by noting what is new about our age.

The fact itself that there is something new is not new. There has been something new about every age; otherwise we would not be able to distinguish them in history. What we need to examine is what in particular is new about our age, for the new is not less new just because the old was also at one time new.

The mere prominence in our age of science and technology is not strikingly new, either. A veritable explosion of industrial technology gave its name to a whole age two centuries ago, and it is doubtful that any scientific idea will ever again leave an imprint on the world so penetrating and pervasive as did Isaac Newton's a century before that.

It is not clear, finally, that what is new about our age is the rate at which it changes. What partial evidence we have, in the restricted domain of economics, for example, indicates the contrary. The curve of growth, for the hundred years or so that it can be traced, is smooth, and will not support claims of explosive change or discontinuous rise. For the rest, we lack the stability of concept, the precision of intellectual method and the necessary data to make any reliable statements about the rate of social change in general.

I would therefore hold suspect all argument that purports to show that novelty is new with us, or that major scientific and technological influences are new with us, or that rapidity of social change is new with us. Such assertions, I think, derive more from revolutionary fervor and the wish to persuade than from tested knowledge and the desire to instruct.

Yet there is clearly something new, and its implications are important. I think our age is different from all previous ages in two major respects: first, we dispose, in absolute terms, of a staggering amount of physical power; second, and most important, we are beginning to think and act in conscious realization of that fact. We are therefore the first age that can aspire to be free of the tyranny of physical nature that has plagued man since his beginnings.

The Traditional Tyranny of Matter

The consciousness of physical impossibility has had a long and depressing history. One might speculate that it began with early man's awe of the bruteness and recalcitrance of nature. Earth, air, fire and water—the eternal, immutable elements of ancient physics—imposed their requirements on men, dwarfed them, outlived them and remained indifferent when not downright hostile to them. The physical world loomed large in the affairs of men, and men were impotent against it. Homer celebrated this fact by investing nature with gods, and the earliest philosophers recognized it by erecting each of the natural elements in turn—water, air, earth and fire—into fundamental principles of all existence.

From that day to this, only the language has changed as successive ages encountered and tried to come to terms with physical necessity, with the sheer "rock-bottomness" of nature. It was submitted to as fate in the Athenian drama. It was conceptualized as ignorance by Socrates and as metaphysical matter by his pupils. It was labeled evil by the pre-Christians. It has been exorcised as the devil, damned as flesh or condemned as illicit by the Church. It has been the principle of non-reason in modern philosophy, in the form of John Locke's Substance, as Immanuel Kant's formless manifold, or as Henri Bergson's pure duration. It has conquered the mystic as nirvana, the psyche as the Id, and recent Frenchmen as the blind object of existential commitment.

What men have been saying in all these different ways is that physical nature has seemed to have a structure, almost a will of its own, that has not yielded easily to the designs and purposes of man. It has been a brute thereness, a residual, a sort of ultimate existential stage that allowed, but also limited, the play of thought and action.

It would be difficult to overestimate the consequences of this recalcitrance of the physical on the thinking and outlook of men. They have learned, for most of history, to plan and act *around* a permanent realm of impossibility. Man could travel on the sea, by sail or oar or breaststroke. But he could not travel *in* the sea. He could cross the land on foot, on horseback or by wheel, but he could not fly over it. Legends such as those of Daedalus and Poseidon celebrated in art what men could not aspire to in fact.

Thinking was similarly circumscribed. There were myriad possibilities in existence, but they were not unlimited because they did not include altering the physical structure of existence itself. Man could in principle know all that was possible, once and for all time. What else but this possibility of complete knowledge does Plato name in his idea of the good? The task of thought was to discern and compare and select from among this fixed and eternal realm of possibilities. Its options did not extend beyond it, any more than the chess-player's options extend beyond those allowed by the board and the pieces of his game. There was a natural law, men said, to which all human law was forever subservient, and which fixed the patterns and habits of what was thinkable.

The Promise of Technology

There was occasionally an invention during all this time that did induce a physical change. It thus made something new possible, like adding a pawn to the chess game. New physical possibilities are the result of invention—of technology, as we call it today. That is what "invention" and "technology" mean. Every invention, from the wheel to the rocket, has created new possibilities that did not exist before. But inventions in the past were few, rare, exceptional and marvelous. They were unexpected departures from the norm. They were surprises that societies adjusted to after the fact. They were generally infrequent enough, moreover, so that the adjustments could be made slowly and unconsciously, without radical alteration of world views or of traditional patterns of thought and action. The Industrial Revolution, as we call it, was revolutionary precisely because it ran into attitudes, values and habits of thought and action that were completely unprepared to understand, accept, absorb and change with it.

Today, if I may put it paradoxically, technology is becoming less revolutionary as we recognize and seek after the power that it gives us. Inventions are now many, frequent, planned and increasingly taken for granted. We will not be a bit surprised when we get to the moon. We would, on the contrary, be very surprised if we did not. We are beginning to use invention as a deliberate way to deal with the future, rather than seeing it only as an uncontrolled disrupting of the present. We no longer wait upon invention to occur accidentally. We foster and force it because we see it as a way out of the heretofore inviolable constraints that physical nature has imposed upon us in the past.

Francis Bacon, in the 15th century, was the first to foresee the physical power potential in scientific knowledge. We are the first, as I have suggested, to have enough of that power actually at hand to create new possibilities almost at will. By massive physical changes deliberately induced, we can literally pry new alternatives out of nature. The ancient tyranny of matter has been broken, and we know it. We found, in the 17th century, that the physical world was not at all like what Aristotle had thought and Aquinas had taught. Today we are coming to the further realization that the physical world need not be as it is. We can change it and shape it to suit our purposes.

Technology, in short, has come of age, not merely as technical capability, but as a social phenomenon. We have the power to create new possibilities and the will to do so. By creating new possibilities, we give ourselves more choices. With more choices, we have more opportunities. With more opportunities, we can have more freedom, and with more freedom we can be more human. That, I think, is what is new about our age. We are recognizing that our technical prowess literally bursts with the promise of new freedom, enhanced human dignity, and unfettered aspiration.

Some Dangers and New Problems

At its best, then, technology is nothing if not liberating. Yet many fear it increasingly as enslaving, degrading and destructive of man's most cherished values. It is important to note that this is so, and to try to understand why. I can think of four reasons.

First, we must not ignore the fact that technology does indeed destroy some values. It creates a million possibilities heretofore undreamed of, but it also makes impossible some others heretofore enjoyed. The automobile makes real the legendary foreign land, but it also makes legendary the once real values of the ancient marketplace. Mass production puts Bach and Brueghel in every home, but it also deprives the careful craftsman of a market for the skill and pride he puts into his useful artifact. Modern plumbing destroys the village pump, and modern cities are hostile to the desire to sink roots into and grow upon a piece of land. Some values are unquestionably bygone. To try to restore them is futile, and simply to deplore their loss is sterile. But it is perfectly human to regret them for a time.

Second, technology often reveals what technology has not created: the cost in brutalized human labor, for example, of the few oases of past civilizations whose values only a small elite could enjoy. Communications now reveal the hidden and make the secret public. Transportation displays the better to those whose lot has been the worse. Increasing productivity buys more education, so that more people read and learn and compare and hope and are unsatisfied. Thus technology often seems the final straw, when it is only illuminating rather than adding to the human burden.

Third, technology might be deemed an evil because evil is unquestionably potential in it. We can explore the heavens with it or destroy the world. We can cure disease or poison entire populations. We can free enslaved millions or enslave millions more. Technology spells only possibility, and is in that respect neutral. The new opportunities it gives us include new opportunities to make mistakes. Its massive power can lead to massive error so efficiently perpetrated as to be well-nigh irreversible. Technology is clearly not synonymous with the good. It *can* lead to evil.

Finally, and in a sense most revealing, technology is upsetting because it complicates the world. This is a vague concern, hard to pin down, but I think it is a real one. The new alternatives that technology creates require effort to examine, understand and evaluate them. We are offered more choices, which makes choosing more difficult. We are faced with the need to change, which upsets routines, inhibits reliance on habit and calls for personal readjustments to more flexible postures. We face dangers that call for constant reexamination of values and a readiness to abandon old commitments for new ones more adequate to changing experience. The whole business of living seems to become harder.

The Mistrust of Technology

This negative face of technology is sometimes confused with the whole of it. It can then cloud the understanding in two respects that are worth noting. It can lead to a generalized distrust of the power and works of the human mind by erecting a false dichotomy between the modern scientific and technological enterprises on the one hand, and some idealized and static pre-scientific conception of human values on the other. It can also color discussion of some important contemporary issues, developing from the impact of technology on society, in a way that obscures rather than enhances understanding, and that therefore inhibits rather than facilitates the social action necessary to resolve them.

Because the confusions and discomfort attendant on technology are more immediate and therefore sometimes loom larger than its power and its promise, technology appears to some an alien and hostile trespasser upon the human scene. It thus seems indistinguishable from that other, older, alien and hostile trespasser: the ultimate and unbreachable physical necessity of which I have spoken. Then, since habit dies hard, there occurs one of those curious inversions of the imagination that are not unknown to history. Our newfound control over nature is seen as but the latest form of the tyranny of nature. The knowledge and therefore the mastery of the physical world that we have gained, the tools that we have hewed from nature and the human wonders we are building into her are themselves feared as rampant, uncontrollable, impersonal technique that must surely, we are told, end by robbing us of our livelihood, our freedom and our humanity.

It is not an unfamiliar syndrome. It is reminiscent of the long-time prisoner who may shrink from the responsibility of freedom in preference for the false security of his accustomed cell. It is reminiscent even more of Socrates, who asked about that other prisoner, in the cave of ignorance, whether his eyes would not ache if he were forced to look upon the light of knowledge, "so that he would try to escape and turn back to the things which he could see distinctly, convinced that they really were clearer than these other objects now being shown to him". Is it so different a form of escapism from that, to ascribe impersonality and hostility to the knowledge and the tools that can free us finally from the age-long impersonality and hostility of a recalcitrant physical nature?

Technology has *two* faces: one that is full of promise and one that can discourage and defeat us. The freedom that our power implies from the traditional tyranny of matter—from the evil we have known—carries with it the added responsibility and burden of learning to deal with matter and to blunt the evil, along with all the other problems we have always had to deal with. That is another way of saying that more power and more choice and more freedom require more wisdom if they are to add up to more humanity. The malaise of our age, as many have noted, is that our power increases faster than our ability to understand it and to use it well. But that, surely, is a challenge to be wise, not an invitation to despair.

Technology and Work

An attitude of despair can also, as I have suggested, color particular understandings of particular problems and thus obstruct intelligent action. I think, for example, that it has distorted the public debate about the effects of technology on work and employment.

The problem has persistently taken the form of fear that machines will put people permanently out of work. That fear has prevented recognition of a distinction between two fundamentally different questions. The first is a question of economic analysis and economic and manpower policy about which a great deal is known, which is susceptible to analysis by well-developed and rigorous methods, and on the dimensions and implications of which there is a very high degree of consensus among the professionally competent.

That consensus is that the level of employment is a function of overall economic growth rather than principally of the mechanization and automation of production; that there is little prospect of enforced leisure for large segments of the work force; that there is no evidence either of a growing group of

unemployables in the society, or of a progressive impoverish-
ment of the labor force resulting from the increasing efficiency
of machines.

There are, of course, and there always will be, serious problems for particular individuals, particular occupations and particular industries resulting from shifts in machines and skills as
economic and technological changes continue to occur. But
the theoretical and empirical knowledge, the policy instruments
and the social mechanisms to deal with such problems are better
developed than in any comparable field of social action. They
include measures to signal and cushion the shocks of transition
that are inevitable in a changing economy, no matter how high
and consistent its overall rate of growth. Where action still falls
short of need in this field, it is for lack of political will, not
knowledge, that it does so.

There is not much that is significantly new, in other words,
in the probable consequences of automation on employment.
Automation is but the latest form of mechanization, and it has
been recognized as an important factor in economic change at
least since the Industrial Revolution. What *is* new is a height-
ened social awareness of the implications of machines for men,
which derives from the unprecedented scale, prevalence and
visibility of modern technological innovation. That is the second question. It also is a question of work, to be sure, but it is
not one of employment in the economic connotation of the
term. It is a distinct question that has been too often confused
with the economic one because it has been formulated, incorrectly, as a question of automation and employment.

This question is much less a question of whether people will
be employed than of what they can most usefully do, given the
broader range of choices that technology can make available
to them. It is less a technical economic question than a question of the values and quality of work. It is not a question of
what to do with increasing leisure, but of how to define new occupations that combine social utility and personal satisfaction.

I see no evidence, in other words, that society will need less work done on some day in the future when machines may be largely satisfying its material needs, or that it will not value and reward that work. But we are still a long way from that day, so long as there remain societies less affluent than the most affluent. Moreover, there is work of education, integration, creation and eradication of disease and discontent to do that is barely tapped so long as most people must labor to produce the goods that we consume. The more machines can take over what we do, the more we can do what machines cannot do. That, too, is liberation: the liberation of history's slaves, finally to be people.

That, I think, is the real meaning of technology for work. To claim the contrary, that machines will impoverish and dehumanize us, is to stand our time on its head and to misread the evidence before us. It is to abandon reason for despair and argument for cant, and it prompts me to recall, for a new generation, Franklin Roosevelt's insight that "the only thing we have to fear is fear itself".

The Fear of Science

Such basically irrational fears of technology have a counterpart in popular fears of science itself. Here, too, anticipatory despair in the face of some genuine problems posed by science and technology can cloud the understanding.

It is admittedly horrible, for example, to contemplate the unintentional evil implicit in the ignorance and fallibility of man as he strives to control his environment and improve his lot. What untoward effects might our grandchildren suffer from the drugs that cure our ills today? What monsters might we breed unwittingly while we are learning to manipulate the genetic code? What are the tensions on the human psyche of a cold and rapid automated world? What political disaster do we court by providing 1984's Big Brother with all the tools he will ever need? Better perhaps, in Hamlet's words, to

> . . . bear those ills we have
> Than fly to others that we know not of.

Why not stop it all? Stop automation! Stop tampering with life and heredity! Stop the senseless race to the moon! The cry is an old one. It was first heard, no doubt, when the wheel was invented. The technologies of the bomb, the automobile, the spinning jenny, gunpowder, printing—all provoked social dislocation accompanied by similar cries of "Stop". Well, but why not stop now, while there may still be a minute left before the clock strikes twelve?

We do not stop, I think, for three reasons: we do not want to, we cannot and still be men, and we should not.

It is not at all clear that atom bombs will kill more people than wars have ever done, but energy from the atom might one day erase the frightening gap between the more and less favored peoples of the world. Was it more tragic to infect a hundred children with a faulty polio vaccine than to have allowed the scourge free reign forever? It is not clear that the monster that the laboratory may create, in searching the secret of life, will be more monstrous than those that nature will produce unaided if its secrets remain forever hidden. Is it really clear that rampant multiplication is a better ultimate fate for man than to suffer, but eventually survive, the mistakes that go with learning? The first reason we do not stop is that I do not think we would decide, on close examination, that we really want to.

The second reason is that we cannot if we would retain our humanity. Aristotle saw a long time ago that "man by nature desires to know". He will probe and learn all that his curiosity prompts him to and his brain allows, so long as there is life within him. The stoppers of the past have always lost in the end.

We do not stop, finally, because we would not stop being men. I do not believe that even those who decry science the loudest would willingly concede that the race has now been proved incapable of coping with its own creations. That admission would be the ultimate in dehumanization, for it would be to surrender the very qualities of intelligence, courage, vision and aspiration that make us human. "Stop", in the end, is the last desperate cry of the man who abandons man because he is defeated by the responsibility of being human. It is the final failure of nerve.

The Recovery of Nerve

I am recalling that celebrated phrase, "the failure of nerve", in order to introduce a third and final example of how fear and pessimism can color understanding and confuse our values. It is the example of those who see the sin of pride in man's confident mastery of nature.

The phrase, "the failure of nerve", was first used by the eminent classical scholar, Gilbert Murray, to characterize the change of temper that occurred in Hellenistic civilization at the turn of our era. The Greeks of the 5th and 4th centuries B.C. believed in the ultimate intelligibility of the universe. There was nothing in the nature of existence or of man that was inherently unknowable. They accordingly believed also in the power of the human intelligence to know all there was to know about the world and to guide man's career in it.

The wars and mixing of cultures that marked the subsequent period brought with them vicissitude and uncertainty that shook this classic faith in the intelligibility of the world and in the capacity of men to know and to do. There was henceforth to be a realm of knowledge and action available only to God, not subject to reason or to human effort. Men, in other words, more and more turned to God to do for them what they no longer felt confident to do for themselves. That was the failure of nerve.

The burden of what I have been saying thus far is that times are changing. We have the power and will to probe and change physical nature, to control our own biology and that of the animals and plants in our environment, to modify our weather, to alter human personality, to reach the moon today and the rest of the heavens tomorrow. No longer are God, the human soul or the mysteries of life improper objects of inquiry. We are ready to probe whatever our imagination prompts us to. We are convinced again, for the first time since the Greeks, of the essential intelligibility of the universe: there is nothing in it that is in principle not knowable. As the sociologist Daniel Bell has put it: "Today we feel that there are no inherent secrets in the universe, and this is one of the significant changes in the modern moral temper." That is another way of stating what is new about our age. We are witnessing a widespread recovery of nerve.

Is this confidence a sin? "Probably most Christians," according to Gilbert Murray, "are inclined to believe that without some failure and sense of failure, without a contrite heart and conviction of sin, man can hardly attain the religious life." I would suspect that this statement is still true of most Christians, although it is clear from the recent literature that a number of theologians are coming to a different view. I think myself that to see a sense of failure toward the world as a condition of religious experience is an historical relic, dating from a time when an indifferent nature and a hostile world so overwhelmed men that they gave up thought for consolation. To persist in such a view today, when nature is coming increasingly under control as a result of restored human confidence and power, is both to distort reality and to sell religion short. It surely does no glory to God to rest his power on the impotence of man in dominating the world.

The Relevance of Religion

Such misconstruings of the nature of our age as I have been describing—whether of its economy, its science or its temper—threaten the religious enterprise by blinding it to the new role it must define for itself if it is to remain effective in the world. They come from men who have yet to recover their nerve—or who have lost it anew—and who therefore see only evil in power, danger in knowledge and sin in human confidence. They thereby obscure the crucial question that technology raises for religion: the question of its relevance. Religion is in crisis today because it has largely lost its old role and has yet to find its new one. "After twenty centuries of doing man's work," I once wrote in a parenthesis, "the Churches are now having to learn to do God's."

What is man's work, and what is God's? Man's work is to be wise and good. It is God's work to reveal himself to man as wisdom and as goodness. It is man's work to discern value and to realize possibility. It is God's to be ultimate value and eternal possibility, and to imbue man with the grace to know and to worship him. It is man's work to know God, and God's to be knowable by man. In other words, God and man are partners in the work of the world, which means at least that man must do his part.

But his part is precisely what man has not done for most of the centuries since the time that Murray wrote about. Abandonment of the belief in intelligibility 2,000 years ago was justly described as a failure of nerve because it was the prelude to moral surrender. Men gave up the effort to be wise because they found it too hard. They gave up the hope of bliss in this world, saw value only in the next, and attributed power only to God. They thus left their work undone. The Churches, if only by default at first, sought to fill the void and do it for them by doing their knowing, their building, their ruling and their moral suasion. Opinions differ and disputes have raged about how well or ill they did, but there was never a question of their relevance to men's concerns, for it was men's concerns they made their own. The Churches were doing man's work.

Renewed belief in intelligibility 2,000 years later means that man takes up his own burden once again. That, as I have tried to show, is what is really new about our age. It means that the Churches are out of man's work, because man now chooses to do his own and has the knowledge and the skill and the will to do it. It is idle to tell him that he cannot because he is ignorant or weak or a sinner. He will not listen.

Does renewed faith in intelligibility also mean, therefore, that the Churches are completely out of work? They have served, throughout their history, as man's crutch. Is their history to end, now that man can walk again? The crisis of the Churches is in the fear of irrelevance born of man's recovery of nerve. It is an agonizing crisis, because there is nothing so dead as irrelevance, which leaves not even a memory.

Knowing God and Doing His Work

The thought with which I would leave you is that the resolution of this crisis may lie in the very power and temper of mind that have brought it about. The point has been prefigured in my text. It is that man's newfound power and confidence enable him to pick once more his partnership with God in doing the work of the world. His need to know God is therefore relatively greater now than it was in an earlier, more frightened time, when just to trust him seemed enough. By the same token, the Churches are freed of the burden of doing man's work and may find their new vocation in doing God's: in knowing God and showing him to man. I say their "new vocation" because the Churches have very often in the past been too busy with man's work to give their undivided attention to God's. And I say they "may find" it because they may not. If they do not, someone else will. It is occasionally sobering to recall that, while God is eternal, the Churches are not, and that even the oldest among them are but recent additions to the immemorial span of human religious experience.

What does it mean to know God? It means, in St. Paul's terms, to look "not at the things which are seen, but at the things which are not seen: for the things which are seen are temporal; but the things which are not seen are eternal".

I have said that technology gives man the power to create new possibilities and the will to do so. As he exercises that power and will, he encounters the problem of how to see and choose the good amid all that is possible. I have also suggested that God reveals himself as goodness and as eternal possibility. That is why man's need to know God grows, the more competent he himself grows in the performance of his work.

And what does it mean to show God to man? I take my text here from Professor Shinn:

> Christian faith has always been concerned with the meeting of the ultimate with the concrete. It can be content neither with visions of eternity nor with purely pragmatic judgments that see decisions only in their immediate contexts. . . . The Church knows a Gospel, which constantly reminds it that the test of any social proposal is not loyalty to treasured traditions but concern for the welfare of persons who are loved by their creator and redeemer. . . . A major part of its calling in our day is to keep questioning society—asking for the purposes, the values, the effect upon persons of the processes the society takes for granted.

There seem to me to be three indispensable prerequisites for the Churches to perform that office. First, the Churches must be of—but stand a bit apart from—the society. I occasionally detect a danger of oversecularization of the Church herself, which would put her back into man's work again, only alongside rather than just for him this time.

Second, the Churches must stand a bit apart from society, but not be wholly sundered from it. The Arian heresy was justly condemned, and it is neither less heretical nor more helpful in some of its present forms.

Third, if they would criticize society constructively, the Churches must above all know whereof they speak. They must know the society's knowledge, appreciate its power and understand its aspirations. They must show God to man as he is, not as they would have him be, for that would be blasphemy.

Eric Mascall/*London, England*

The Scientific Outlook and the Christian Message

THE BASIC FACT

The basic fact of the situation which confronts the Church today is, I suggest, that the enormous domination of present-day life by scientific technology has produced a psychological climate in which it no longer comes naturally to people to attend to those aspects of the world which manifest it as the creature of God. Our minds have been conditioned to look upon the world as raw material for human manipulation rather than to contemplate it in an attitude of wonder. Not "What is it?" or "Why is it?" is the question that springs to our minds, but "What can we (or the technologists) do with it?" When we do wonder, it is at the skill of the technologists rather than at the creativity of God.

Scientists themselves do ask the questions "What is it?" and "Why is it?", even if only on an empirical level, and they are on the whole modest and sometimes devout men, very conscious of the terrible potentialities of the forces that are in their hands. We are not, however, now thinking of scientists themselves, or at any rate not of scientists as such, but of the great multitude of people who are not scientists but who have grown up and live in a world that scientific technology has built up and dominates. To them the realities of this world are the only realities that there are, and the scientist and the technician are the masters of their secrets.

II

THE SECULAR OUTLOOK OF MODERN MAN

Some years ago a great chemical firm published an advertisement which began by reminding readers that in the ancient world wizards were held in great awe because of the changes they were able to produce, which were often unpleasant ones. "The wizard of the present day," it continued, "is the analytical chemist, but he is always welcome, since the changes which he produces are always for the better." Apart from its naive conclusion, this sentence provides a fairly accurate statement of the place that the applied scientist has come to hold in the contemporary mind.

Two other factors in the situation strengthen this outlook; they are factors of very different kinds.

In the first place, most of the this-worldly needs for which Christians used to pray to God appear to be under the control of scientific experts, or, if not yet under their control, predictable by them. We no longer pray to God for recovery from sickness; we take refuge in the medical services. We no longer pray for rain; we read the weather forecasts, and we feel pretty confident that before very long the meteorologists will be able not only to predict the weather but also to control it.

Secondly, the modern media of publicity and the vast advertising industry which employs them combine to instill into our minds an entirely this-worldly scale of values, which at best is that of a purely naturalistic morality and at worst is, even by naturalistic standards, base and shoddy.

Furthermore, we must remember that it is of the essence of science, and also the secret of its success, that it attends precisely to those aspects of reality which are, at least relatively, deterministic. Biologists indeed deal frequently with statistical aggregates whose behavior can be only probably predicted; however, this does not imply that the individual elements are undetermined but only that we are ignorant of their details or interested in them only in the mass.

Physicists—those of the "Copenhagen school"—hold that the basic submicroscopic events are undetermined, but this does not imply any effective indeterminism in the behavior of the aggregates perceptible to our senses. It is, of course, true that science has been so amazingly successful because it has attended to those aspects of reality—the measurable and predictable aspects —which are amenable to its method; but when it has been so successful it is hardly surprising, even if it is rationally unjustified, that we tend to take it for granted that all aspects of reality will be ultimately brought within the scope of scientific determinism or, at any rate, that those which cannot be brought are of no real importance.

Thus, in the face of the plain deliverances of our own consciousness, we tend to deny the freedom of our wills and the whole notion of responsibility; as a result, our very natural (if fallenly natural) tendency to conform to our environment and copy our neighbors receives a pseudo-confirmation from the sciences of psychology and endocrinology. If my actions are determined by my complexes and my glands, how can I be really free in willing and acting? It requires more sophistication than most people possess to recognize that a consistently deterministic view of human behavior saws off the branch on which the psychological determinist is sitting. What if the psychologist's belief in his system is only the result of his own psychological conditioning? [1]

Nevertheless, the practical outcome is that, in place of the personal notions of divine providence and human freedom, we find ourselves overshadowed by the impersonal notions of necessity and chance (for chance, as another name for our ignorance

[1] For a reply to attempts to avoid the dilemma that psychological determinism is either false or self-refuting, cf. E. L. Mascall, *Christian Theology and Natural Science* (London, 1956), pp. 212ff.

of the precise functioning of necessity, offers a specious, though spurious, escape from it). The vast growth of gambling in Britain in recent years is one obvious manifestation of this; like the ancient Jews, those who forget the Lord and forsake his holy mountain set a table for Fortune and fill up cups of mixed wine for Destiny.[2] But we may remember that in the ancient world necessity and chance, Destiny and Fortune, *anagke* and *tyche,* were two of the bogeys from which Christianity appeared to man as a deliverer.

III

COMPATIBILITY OF SCIENCE AND CHRISTIAN DOCTRINE

I do not think it can be validly maintained that the secularist outlook of modern man is entirely to be attributed to science, but it is clear that in several ways science has played a very large part, perhaps the major part, in producing it. There is, first of all, the sheer staggering success of scientific technology in transforming the world and man's life in it, taken with the fact that, by its very method, science abstracts from those aspects of the world which can lead men to see it as the creature of God.

In itself this is nothing for the Christian to lament; indeed, with Teilhard de Chardin, he may rightly rejoice that man, God's vicegerent on earth, has been allowed to penetrate so deeply into the secrets of God's world and to bring it under his control. In its right place man's concern with "this world" is legitimate and indeed inevitable, for in his physical constitution he is part of it and his continuing existence depends on his ability to exploit it without destroying it. Nevertheless, the dangers that such an explosive expansion in the realm of the secular offers to a fallen being are obvious; he may forget both that this world is God's creature and that his own destiny, while bound up with it, lies beyond it.

[2] Is. 65, 11 (RSV).

The fragmentation and debility of Western Christendom at the time when the scientific explosion began made the Church almost entirely incapable of coping with it, and in spite of the real religious devotion of many of its leading figures, it rapidly escaped from religious control and direction. Side by side with this, and largely bound up with it, was the growth of economic capitalism with its constant need to produce and to sell more and more material goods in order to survive. Lastly, the "shrinkage of the earth" in consequence of the vastly increased speed of travel and the development of the media of communication—themselves the products of science—have made the dissemination of the basic ideas of a secularist culture—indeed a very degraded type of secularist culture—more and more easy and insidious, so that the ideas and ideals that the means of communication instill into us are largely devised by people who want to make us buy what they have to sell.

There is, of course, still the possibility for the discerning and sophisticated person to make his own enclave and for cultural efforts to flourish with some degree of independence. But the great body of men and women for whom Christ died are neither discerning nor sophisticated, and even the sophisticated enclaves are more influenced by their environment than they are usually ready to recognize. Elites are inevitably parasitic upon the general cultural structure of the community, and since all men are equal in the sight of God, it is a healthy thing that they should be.

There is still, I believe, a need and a place for the old kind of apologetic which consists in arguing that science and Christian doctrine are not in fact incompatible. This should include the frank recognition that scientific theories are never final and also that Christian doctrine, while in its essence unchanging, may in its expression, emphasis and detail vary a great deal from one age to another, and indeed that the new knowledge of the world which comes to us from science, and even the special interests of a secularist society, may provide the stimulus for the development of doctrine. There is nothing for Christian theology to fear from this, as long as it remembers that all human knowledge is necessarily incomplete, and as long as it avoids both frightened denunciations and premature syntheses.

With these qualifications, it is useful to note that theology of the traditional Catholic type, just because it has a positive doctrine about the created world, is especially equipped to explore those regions where its subject matter overlaps that of science. I have argued elsewhere, to give one example, that the possibility of intelligent corporeal life in other parts of the universe than this earth of ours is fully compatible with the orthodox Chalcedonian christology but is much more difficult to reconcile with the "kenotic" views.[3]

Nevertheless, important as these detailed discussions are from the point of view of the Church's apostolate to intellectuals, it is far more urgent that we shall find how to speak to the great mass of men and women who, while having little or no understanding of science itself, are conditioned by the atmosphere and the assumptions of a technocratic civilization. It is the desire to find a point of contact here that has led a number of Christian thinkers to produce the various types of secularized Christianity which accept the secularized world at its own valuation and remodel the Christian faith to accord with it.

IV

TWO DEFECTS OF SECULARIZED CHRISTIANITY

This method has, as I see it, two glaring defects. First, it only too easily results in changing not only the expression but the essence of Christianity and in reducing it simply to secularism with some trappings of Christian language. Second, just because it accepts the world's valuation of itself, it has no criteria by which it can discriminate between the true and the false elements in that valuation. In spite of its talk about the

[3] E. Mascall, *op. cit.,* pp. 36ff.

necessity for the Church to die, it is really a program for the Church to survive when she has lost her *raison d'être*.

I have examined at length elsewhere some of these systems of secularized Christianity[4] and I cannot go into more detail here. What, I believe, is urgently needed today is the development of a theology of the secular, a *théologie des réalités terrestres,* that is, a constructive discussion of the concerns of the contemporary world in the light of the Christian doctrine of God, the world and man. This is a task that cannot be carried out by theologians alone; it needs the frank and sympathetic collaboration of experts in the various fields of contemporary thought and activity, medical men, scientists, sociologists, lawyers, educationists, welfare workers, industrialists and the rest. And the purpose of such collaboration will not simply be to help the theologian to produce better and more relevant theology, though this should be one of its results; the purpose will rather be to enable the Church to get to grips more intelligently and effectually with the problems of the contemporary world—and this is a predominantly lay activity.

V

DIALOGUE WITH THE CONTEMPORARY WORLD

The potentialities, both for good and for evil, of our science-based technological culture are immense. The prescientific era might be described as one in which man was under the control (and also, in a sense, as Karl Rahner has pointed out,[5] under the protection) of nature. The scientific era might be described as one in which nature has been more and more brought under the control of man. Now, however, we are moving into an age in which, by his insight not only into the world of nature around him but also into himself, man will be able to produce radical changes in his own being.

[4] *The Secularization of Christianity* (London, 1965).
[5] K. Rahner, *Mission and Grace* III (London, 1966), pp. 22ff.

The recent work in such sciences as psychology, pharmacology, neurophysiology, molecular biology and genetics provides the proof of this. The ethical problems that arise from these advances are puzzling to the Christian moralist, but they are not less puzzling to the secular humanist. If man is the measure of all things, by what is man himself to be measured? The Christian doctrine of man may need to be developed and deepened, but at least the Christian has a doctrine, and it is one which, by asserting that man is subject to God, makes everything else subject to man. Its further assertions—that God has himself taken human nature in Christ and that our human nature can be renewed by incorporation into him—give added strength to its affirmation of the indefeasible value and dignity of man. In a plural society, such as that in which most Christians live today, it may be at this point that they can make contact and enter into dialogue with their humanist neighbors.

Finally I would stress the importance, in our dialogue with the contemporary world, of the doctrine of the resurrection of the body. Since by our bodily constitution we are quite literally part of the material world, this doctrine entails nothing less than the ultimate transfiguration of the material world and its absorption into Christ. St. Paul was quite clear that the redemption of man involves the eschatological transformation of the whole of creation.[6] Only too often, however, Christians have fallen into a virtually Manichaean spiritualism which sees man's beatitude as consisting in a complete break with the realm of matter and a definitive escape from bodily existence.

Teilhard de Chardin has done perhaps more than any Christian writer in recent years to recall us from this error; however much his exposition may need correcting and supplementing, this is what his doctrine of the Omega point means. The orthodox Christian, then, is concerned not less than the secularist with the fate and the vicissitudes of "this world", and he may indeed validly claim that, if words are used correctly, Christianity is the most radical and authentic humanism. For it holds that

[6] Rom. 8, 22.

man—not merely his soul—is made for the vision of God, and that the kingdoms of this world are to become the kingdom of God and of Christ.[7] Furthermore, the involvement of the Christian in the life of the world will be seen to be nothing less than his participation as Christ's agent and instrument in this work of transfiguration, a work which is going on here and now, since grace is the beginning of glory.[8]

It is, I suggest, at this point of a common concern with the secular order that the Christian may hope to enter into dialogue with his fellowmen in this technocratic culture in which we all live, and his conviction that the secular order has a destiny beyond itself will lead him to value it more, and not less, than it is valued by the secularist.

[7] Apoc. 11, 15.
[8] Cf. Thomas Aquinas, *Summa Theologica* II-II, 24, 3 ad 1.

PART II
BIBLIOGRAPHICAL
SURVEY

Ben van Onna/*Tübingen, W. Germany*

The State of Paradise and Evolution

To start with, Christian teaching is only indirectly interested in the state of paradise. Only in the perspective of redemption from sin can we get a correct view about the meaning and content of a preternatural condition connected with the original situation of the first human beings.[1] It is therefore not without reason that both scripture and the ecclesiastical magisterium[2] speak so cautiously and indirectly about paradise and Adam.

On this rather narrow basis theology built up a teaching in the course of its history[3] which has been subjected to increasing

[1] Cf. C. Dumont, "La prédication du péché originel," in *Nouv. Rev. Théol.* 83 (1961), pp. 113-34; P. Schoonenberg, *Het geloof van ons doopsel, IV: De macht der zonde* ('s Hertogenbosch, 1962), esp. pp. 5-12, 73-5; 272-80; P. Smulders, *Het visioen van Teilhard de Chardin* (Bruges, ³1963), pp. 236-42; H. Fiolet, "De erfzonde als verbonds-mysterie," in *Jaarboek 1963/4 Werkgenootschap van katholieke Theologen in Nederland* (Hilversum, 1965), pp. 72-85.

[2] Cf. A. Vanneste, "La préhistoire du décret du Concile de Trente sur le péché originel," *Nouv. Rev. Théol.* 86 (1964), pp. 355-68, 490-510; *idem*, "Le Décret du Concile de Trente sur le péché originel," *ibid.* 87 (1965), pp. 688-726; K. Rahner, "Theologisches zum Monogenismus," in *Schriften zur Theologie* I (⁶1962), pp. 255-75; see also footnote 3.

[3] Cf. A. Gaudel, "Péché originel," in *Dict. Théol. cath.* XII (1933), pp. 317-582; J. Gross, *Geschichte des Erbsündendogmas, I: Entstehungsgeschichte des Erbsündendogmas. Von der Bibel bis Augustinus* (Munich, 1960): *II: Entwicklungsgeschichte des Erbsündendogmas.* (Munich, 1963); P. Schoonenberg, *op. cit.*, pp. 117-42, 155-82; J. Hasselaar,

criticism on the part of philosophy, evolutionism (particularly with regard to man) and recently also of exegesis.

This article seeks to survey present Catholic theology about paradise[4] insofar as it has been affected by science and the consequent evolutionary image of the world. Here the first question concerns the historical aspect of paradise before the fall. Since I want primarily to indicate various *theological* interpretations of the situation at man's origin, I shall refer *en passant* to some points that have come to light in biblical exegesis.

1. *Dissatisfaction with Current Teaching*

Present-day dissatisfaction with traditional teaching centers on the following difficulty. Contemporary sciences (paleontology, evolution and biology) present us with a picture of primitive man essentially different from that "superman" of paradise presented at an early stage by Christian theology under the influence of late Jewish apocalyptic and Hellenistic thought.[5] Thus the first historical human being, Adam, was endowed with a perfection which could be inherited and which surpassed by far the potentiality of all his descendants.

In the evolutionary perspective everything begins as something small, primitive and imperfect, and this also holds for the origin of man at the moment he rises successfully above the

---

Erfzonde en vrijheid (The Hague, 1953). A more recent study of St. Thomas on original sin and paradise may be found in W. Van Roo, *Grace and Original Justice according to St. Thomas* (Rome, 1955).

[4] The various views on the original state of man among Protestant theologians since the 18th century are indicated in P. Wrzecionko, "Urstand, IV" in *Rel. in Gesch. u. Gegenw.* VI ([3]1962), pp. 1208-12. A short summary of modern Protestant opinion on paradise has been given by E. Jacobs, "Urstand," in *Evang. Kirchenlexikon* III ([3]1959), pp. 1597-600; O. Michel, P. Wrzecionko and K. Schlink, "Urstand," in *Rel. in Gesch. u. Gegenwart* VI ([3]1962), pp. 1205-14. Starting from existential dialectics these theologians (esp. K. Barth and E. Brunner) consider the question of the historicity of Adam and paradise as disposed of. And so they have been less preoccupied about the border between science and theology in relation to the question of man's original situation.

[5] Cf. J. Jeremias, "Paradeisos," in *Theol. Wörterb. z. N.T.* V (1954), pp. 863-71; E. Cothenet, "Paradis," in *Dict. Bibl., Suppl.* VI (1960), pp. 1177-220.

primitive forms of animal life.[6] Was an extremely primitive first man really capable of making a decision of the caliber suggested by theology and thus be responsible for a total regression at the beginning of human history? [7]

2. *Earlier Attempts at a Theological Solution*

Theology tried out all kinds of ways in order to escape from this dilemma between perfection and imperfection. It tried, for instance, to sever theology from ordinary science so that man could bodily be subject to evolution while his soul became exclusively a matter for theology.[8] But even the most sensible theological escape, which maintained that the biological appearance of the first man did not have to coincide with hominization in the theological sense,[9] was not satisfactory. One cannot prize the theological Adam out of the framework of evolutionary history, for in that case this "Adam" would be a "man" with whom we, in our time, would have no connection at all, and who would therefore be meaningless for us, even if we would credit him with special theological significance. Moreover, one cannot debase divine grace, totally realized at man's origin according to traditional theology, to the execution of an extraordinary theatrical display within this world (Rahner) which would be irreconcilable with the actual evolution-conditioned abilities of primitive man.

[6] Cf. A. Hulsbosch, *De schepping Gods, Schepping, Zonde en Verlossing in het evolutionistische wereldbeeld* (Roermond, 1963), pp. 25-57; P. Smulders, *op. cit.*, pp. 233, 266-7; P. Schoonenberg, *Gods wordende wereld. Vijf theologische essays* (Woord en Beleving, 13) (Tielt, 1962), pp. 9-60; for the more traditional views see J. Feiner, "Oorsprong, oerstaat en oergeschiedenis van de manes," in *Theologisch Perspectief* II (Hilversum, 1959), pp. 29-60; F. M. Bergounioux and P. Hermand, "Homme," in *Catholicisme* V (1962), pp. 843-62; M. M. Labourdette, *De erfzonde en de oorsprong van de mens* (Voorhout, 1956), pp. 37-53; B. Piault, *La création et le péché originel* (Paris, 1960), pp. 134-49.

[7] S. Trooster, *Evolutie in de erfzondeleer* (Bruges, 1965), p. 25.

[8] Cf. the important contribution by K. Rahner, *Die Hominisation als theologische Frage;* P. Overhage and K. Rahner, *Das Problem der Hominisation. Über den biologischen Ursprung des Menschen* (Quaestiones Disputatae 12 and 13) (Freiburg, ²1963), pp. 13-90.

[9] P. Smulders, *op. cit.*, p. 273; in many studies (e.g., J. Feiner, *art. cit.*, F. M. Bergounioux and P. Hermand, *art. cit.*) this possibility is simply taken for granted.

The same can be said of those concrete analyses of man's original perfection that are common in the theology of paradise. These specific gifts can only be accepted if we eliminate from paradise those laws of mortality and concupiscence which prevail in human history after the fall.[10] In that case human nature was something very different in primitive man from what it is for us today and in no way corresponded to that existence which preceded man.[11]

However, death and the misery which accompanies it are conditions basic to the situation and processes of nature and the universe. Nor can one imagine a world without concupiscence, without that arbitrary lust which precedes the action of man's free will and does not let itself be harnessed by this will without asserting itself. If we cut out these laws from this world of ours, even if only for a small area of space and time which we call paradise, we shall simply be left with a totally unreal man, particularly if this man has to be the Adam of paradise who must have been wholly bound up still with his prehuman condition. In other words, it is precisely those conditions which cannot be reconciled with the primitive situation and which in such a theory become the very condition of the existence of man, of life and of the order of the world.[12]

Finally the difficulties of this traditional theology of paradise are underlined by our modern sense of history. It is feared that this view of paradise emasculates all genuine history in that an ideal of original perfection is made the standard by which to judge every historical development. Thus the history of humanity which shows a regression from man's origin after the fall would then have no other purpose than to return in a vast circu-

[10] For some time theology has no longer been interested in a third original preternatural gift, namely the exceptional knowledge given to Adam, because this is clearly based on a wrong interpretation of Gen. 2, 20 and 24.

[11] See the literature mentioned in footnote 3 and the bibliographical survey in H. Haag, *Biblische Schöpfungslehre und kirchliche Erbsündenlehre* (Stuttgarter Bibelstudien, 10) (Stuttgart, 1966), pp. 13-40.

[12] See H. Renckens, *Israels visie op het verleden. Over Gennesis 1-3* (Tielt, ⁶1965), p. 148.

lar movement to the perfect origin from which it started.[13] Is such a presentation not influenced by remnants of some myth or out-of-date way of thought, apart from making it very difficult to come to a genuine and biblical theology of creation? [14]

3. *The Exegetical Approach*

Contemporary exegesis has also turned away from this "romantic intercourse with the past" (Lohfink), which seemed to be indicated by the literal interpretation of Genesis 1-3. The study of this section shows first of all that the origin of this narrative must not be sought in a kind of eye-witness report on the historical events of creation and paradise. Otherwise there would be no need for metaphors and images. Nor does this narrative represent the notation of a tradition handed down from the very beginning—the enormous lapse of time would make this in any case impossible—or the substance of a vision which revealed to the author historical events of millions of years ago. Nor is the text a catechetical presentation of the facts to a primitive audience while the author "knew much better himself".[15]

What we are given in Genesis 1-3 is "Israel's view of the past".[16] There the people of Israel look at things in the light of their vocation and of God's promise for this world, and so they use the past of the world and of mankind in order to interpret that actual history of salvation. If what they are experiencing in their lives is given meaning by grace and judgment, then, according to Israel's way of thinking, it must have been the same right at the beginning of history: on the one hand, God's faithfulness as the beginning of the promise manifested in the goodness of creation (Gen. 1), on the other, the unfaithfulness of

---

[13] P. Smulders, *op. cit.,* p. 233; B. Delfgaauw, *Geschiedenis en vooruitgang III: De eeuwigheid van de mens* (Baarn, 1964), pp. 168-9.

[14] See A. Hulsbosch, *op. cit.,* esp. pp. 25-7; Hulsbosch views God's creative action no longer from the beginning but from the end.

[15] See H. Renckens, *op. cit.,* pp. 25-40, 121-41; J. Weterman, "Bijbel en evolutie," in *Katholiek Artsenblad* 41 (1962), pp. 25-9; K. Rahner, *art. cit.* (Die Hominisation . . .), pp. 33-5.

[16] This is the title of H. Renckens' generally commendable book, *op. cit.*

the first human beings in the "fall" (Gen. 2-3) and an accelerated sinfulness of all mankind (Gen. 4-11).[17]

Dealing with this as a dogmatic theologian, Karl Rahner
called this formation of belief in Israel "aetiology", i.e., to look
at something that happened in the past as an explanation of
what is experienced in the present. In this he distinguishes between a "mythological" and an "historical" aetiology. In the
first case, the past is simply used as a symbolic description in
order to explain the concrete situation in which man finds himself from one age to another. It is in this way that, according
to Rahner, the Protestants seem to prefer to interpret the narrative of Genesis.[18] Historical aetiology would be at work when
a view of the past fits in rightly and viably with the present—in
other words, when it reveals a historical cause for the present
situation, as happens in the bible.[19]

Now Rahner seems to suggest that the process of belief in
Israel is therefore a kind of logical conclusion drawn from an
historical or mythological premise, but this is hardly possible.
Such a retrospective interpretation can only be accepted if scripture itself interprets its narrative in this way or even demytholo-
gizes it. And this is what happens primarily in Genesis 1 which
brings out the real doctrinal content of the story of paradise of
Genesis 2-3.[20]

There is an important corollary to the right understanding

[17] Apart from Renckens, see H. Haag, op. cit., pp. 41-59; N. Lohfink,
"Die Erzählung vom Sündenfall," in Das Siegeslied am Schilfmeer.
Christliche Auseinandersetzungen mit dem Alten Testament (Frankfurt,
1965), pp. 81-101 (De actualiteit van het Oude Testament, 1966); A. M.
Dubarle, Le péché originel dans l'Ecriture (Lection divina, 20) (Paris,
1958), pp. 39-74; H. van den Bussche, De godsdienstige boodschap van
de oergeschiedenis (Kernen en Facetten, 3) (Tielt, ³1963).

[18] K. Rahner, Die Hominisation . . . , p. 36; see also J. Feiner, art.
cit., pp. 31, 51; cf. supra, footnote 4.

[19] K. Rahner, art. cit., pp. 32-42, 85-90; idem, "Theologische Prinzipien
der Hermeneutik eschatologischer Aussagen," in Schriften zur Theologie
IV (1960), pp. 401-28 (applicable to the "protology"). Rahner's first-
named article has been critically examined by N. Lohfink, "Gen. 2f. als
'geschichtliche Ätiologie'," in Scholastik 38 (1963), pp. 321-34, and L.
Alonso-Schökel, "Motivos sapienciales y de Alianza y Gen. 2-3," in
Biblica 43 (1962), pp. 295-316.

[20] See H. Haag, op. cit., pp. 46-7; H. Renckens, op. cit., p. 204.

of the paradise image of Genesis 2-3. Just as Israel looked from the present to the past, so it looked from the same present to the future which God promised to the whole of mankind through Israel. According to Israel's awareness of its faith, the beginning of time ("protology") and the end of time ("eschatology") correspond to each other, and are interpreted in terms of the same present.[21] It is therefore not astonishing that the prophets picture the definitive life of the new mankind in grandiose and paradisean images. In this context these pictures are in their place at the final fulfillment. As an expression of Israel's belief in the past these images are most probably but a literary symbol of the fact *that* (not: *how*) the historical realization of the promise started at the same time as creation in spite of man's unfaithfulness from the very beginning. There is no need to attribute more historical realism to the original state of the first man on the basis of the paradise story.

But even without taking into account this function of the symbolic description of paradise, one cannot really find in the direct content of the narrative (or the rest of scripture) an immediate reason to introduce mortality, lust and a general deterioration of the natural environment after the fall.[22]

4. *More Recent Attempts at a Solution*

It seems evident that both the evolutionary and historical image of the world and the data of scripture make it difficult to accept unquestioningly a special original state of man as historical reality. Thus theology has started in recent years to look

[21] See J. Jeremias, *art. cit.*; E. Cothenet, *art. cit.*; H. Renckens, *op. cit.*, pp. 135-47; P. de Haes, *De Schepping als heilsmysterie* (Woord en beleving, 11) (Tielt, 1962), pp. 63-72; S. Trooster, *op. cit.*, pp. 63-84, 147-57. For the Adam-Christ relationship in the New Testament see E. Brandenburger, *Adam und Christus. Exegetisch-religionsgeschichtliche Untersuchung zu Röm. 5 (1 Cor. 15)* (WMANT 7) (Neukirchen, 1962); P. Lengsfeld, *Adam und Christus. Die Adam-Christus Typologie in Neuen Testament und ihre dogmatische Verwendung bein M. J. Scheeben und K. Barth* (Koinonia, 9) (Essen, 1965); C. K. Barret, *From First Adam to Last* (London, 1962).

[22] Cf. W. Vollborn, "Das Problem des Todes in Gen. 2 und 3," in *Theol. Liter. Zeit.* 77 (1952), pp. 709-14; L. Rost, "Theologische Grundgedanken zur Urgeschichte," *ibid.* 82 (1957), pp. 321-6.

for a new interpretation of paradise. We may distinguish here three main tendencies, of which the first two explicitly maintain the historicity of paradise.

(a) A first group of what one may call "theologizing" scientists starts with the obvious supposition that man bodily descended from some animal form of life. According to them, these first men, who had only just raised themselves above the animal level, felt themselves disoriented with their new consciousness and freedom in this world. Therefore, they were threatened by internal disharmony and external threats from the environment. In order to preserve these first representatives of mankind God must have provided internal and external circumstances to ensure their maintenance—in other words, a state of paradise. The fall interrupted this special situation, and this meant an evolutionary recession and abandoning these creatures again to the forces of this evolution.[23] The original situation is then usually presented as a kind of state of childlike innocence. How in such a situation one explains psychologically such an all-embracing and catastrophic decision as original sin is not clear.

(b) The second group concentrates rather on a new interpretation of the original gifts of immortality and freedom from concupiscence. Following St. Thomas, Rahner sees the heart of the original sinfulness in the opposition between the spontaneous impulses of human nature on the one hand and the free spiritual self-realization of the person on the other. After the fall man was no longer capable of integrating the spontaneous and *then* independent urges of man into the free decision of his personality.

---

[23] Thus, in various ways, A. Haas, "Naturphilosophische Erwägungen zum Menschenbild des Schöpfungsberichtes und der modernen Abstammungstheorie," in *Scholastik* 33 (1958), pp. 355-75; Ph. Dessauer, "Bemerkungen zum Thema: das erste Menschenpaar in heutiger Sicht," in *Schöpfungsglaube und biologische Entwicklungslehre* (Studien und Berichte der Katholischen Akademie in Bayern, 16) (Würzburg, 1962), pp. 133-70; M. Bruna, in M. Bruna and P. Schoonenberg, "Tweegesprek over het ontstaan der zondigheid," in *Tijdschr. v. Theol.* 4 (1964), pp. 56-7; K. Rahner, *Die Hominisation* . . . , pp. 41-2, leaves room for such a solution.

His first innocence embraced both his natural spontaneity and his freedom, but in a relationship which "at that time" could harmonize natural spontaneity and free, spiritual decision.[24] In terms of this same change of relationship between nature and person, Rahner explains the original immortality. Bodily death was a fact, even in paradise, but it was transcended by a personal surrender of man to God who could overcome all bodily resistance.[25]

Now, this somewhat individualistic view of the nature-person relationship does not seem to correspond to the evolutionary and social reality of man. Moreover, for the person to transcend nature in the state of paradise demands a superhuman quality at the start. Finally, this view, too, starts from an historically real paradise, even though one avoids the concrete details of this condition.[26]

One has to admit that Schoonenberg's presentation is really rather similar, although he modifies Rahner's view by saying that the specific character of the original state consisted in man's unlimited striving toward the eschatological realization of immortality and freedom from concupiscence.[27]

(c) The third group is of the opinion that the first human

[24] K. Rahner, "Zum theologischen Begriff der Konkupiszenz," in *Schriften zur Theologie I* ([6]1962), pp. 377-414; B. Stoeckle, *Die Lehre von der erbsündigen Konkupiszenz in ihrer Bedeutung für das christliche Leibethos* (Ettal, 1954); J. B. Metz, "Konkupiszenz," in *Handb. theol. Grundbegriffe* I (1962), pp. 843-51.

[25] K. Rahner, *Zur Theologie des Todes* (Quaestiones Disputatae, 2) (Freiburg, [4]1958), pp. 31-46; similar considerations had already been published earlier by P. Schoonenberg, "Onze activiteit in het sterven," in *Bijdragen Ned. Jezuieten* 6 (1943-6), pp. 127-46; *idem, Het geloof van ons doopsel I: God, Vader en Schepper* ('s Hertogenbosch, 1955), pp. 177-88; R. Troisfontaines, *Ik sterf niet'* (Woord en beleving, 22) (Tielt, 1966), pp. 174-5; L. Boros, *The Mystery of Death* (Ecclesia) (Bruges, 1965), pp. 125-7; see also the bibliography in H. Haag, *op. cit.,* pp. 13-40.

[26] K. Rahner, "Paradies," in *Lex. Theol. u. Kirche* VIII ([2]1963), pp. 71-2.

[27] P. Schoonenberg, "Natuur en zondeval," in *Tijdschr. v. Theol.* 2 (1962), pp. 173-201; the beginning of an eschatological approach is also brought out in J. Feiner, "Urstand," in *Lex. Theol. u. Kirche* X ([2]1965), pp. 572-4.

beings sinned freely against God at the very beginning of their consciousness. This would make the preternatural gifts exist only negatively, sin causing the absence of them, and in this way there was never any objective realization of the grace of paradise. Original justice then existed only in God's plan for our salvation.[28]

Would this justice have existed if man had not sinned? This seems hardly probable. In both cases, whether man had sinned or not, God would have had to make a spectacular intervention in the reality of history and evolution, by way of deterioration or amelioration, in order to make this intervention fit in with unpredictable human actions.

5. *What Next?*

All these attempts presuppose that God's grace really had to appear from the beginning in original justice. But is it not possible to consider the evolutionary reality, according to the findings of modern science, as in itself already a legitimate effect of creation and as embodying the promise of grace at the beginning of man's existence?[29] In that case we would not need a second stage, a paradise, in order to make it plain that God's grace was operative at the very beginning. The original situation then does not mean a paradise that we *no longer* possess, but the final state, the promised and hoped for fulfillment of this world, which does *not yet* exist, but which has been given a start at creation and historically as a task that man must fulfill.[30] A paradise situation at the beginning of mankind which is both fixed and unworldly simply emasculates the character of our faith as a *promise,* and even makes it impossible.

[28] K. Rahner, *Die Hominisation* . . . , pp. 85-8, among other places; S. Trooster, *op. cit.,* pp. 34-5, 80-1, 147-57.
[29] Cf. the important work of A. Hulsbosch, *op. cit.,* particularly pp. 25-57.
[30] P. Schoonenberg, *Het geloof van ons doopsel IV: De macht der zonde* ('s Hertogenbosch, 1962), p. 9; A. Hulsbosch, *op. cit.*

Norbert Schiffers / *Aachen, W. Germany*

Physics Questions Theology

The modern physicist proceeds with his work in three stages: he makes observations and collects data, he mathematically formulates working hypotheses and he tests his formulae by further observations. In this scientific process, technical and methodological problems in the first and third stages take up so much time and trouble that most physicists give, and even have, the impression that the technique and methodology of experimentation comprise their entire employment. The terms the research worker uses to describe his findings, he uses naively, intuitively, without considering their historical or epistemological derivations.[1]

I

DISCUSSIONS OF "MIRACLE"

This engrossment in methodology and technique explains why many modern physicists are bored by the questions that physics can ask theology. It also explains the eagerness of traditional apologetics to engage in conversation with physicists about the possibility of so-called miracles.

Of course these discussions should have taken particular care

[1] A. Einstein, Foreword to M. Jammer, *Concepts of Space* (Cambridge, 1953), p. xii.

to clarify what the physicist and what the theologian understand by nature. A meeting of the *Paulusgesellschaft*[2] on science and miracles finally made the following points:

1. Scientific observation works with the methodological assumption that the world constitutes a closed and therefore available reality. This concept of nature requires a methodological determinism to explain the world.

2. Although (at any rate since quantum physics) physics is in fact indeterministic—a point made by Johannes Hessen, Heimo Dolch and Wolfgang Büchel [3]—the contingency resulting from this indeterminism is a physical contingency, and if it manifested itself, this would be no miracle in the theological sense, but the natural possibility of an exception.

3. The miracle's "place" is not the laboratory of the physicist, to whom an exception presents difficulties in the working out of his system, but the jungle where man makes his existential search for meaning; if something marvelous happens, he recognizes it as a message from another person. In other words, a miracle is a communicative, finite sign for an unfulfilled faith; as a sign it cannot be questioned by physics, but in its factual content it can. "The gap remains between the establishment of the fact—the empty tomb, for example—which in itself is not concerned with miracle and seeks only to discover whether "the bible is right", and the sign of faith which confronts a man and will not let him rest at a purely natural explanation. The point made in the discussion was simply that nature for the physicist is a methodologically closed system, but in the existential analysis of theological anthropology it remains open to the supernatural. Theology which is not simply apologetics should at least try to show how its concept of nature is complementary to that of the sciences.

[2] *Dokumente der Paulusgesellschaft,* ed. E. Kellner, Vol. 5, "Wunder und Wissenschaft" (Munich, 1963), pp. 190-231, 250, 261.
[3] J. Hessen, *Das Kausalprinzip* (Augsburg, 1928); H. Dolch, *Kausalität im Verstandnis des Theologen und der Begründer der neuzeitlichen* (Freiburg, 1954); W. Büchel, *Philosophische Probleme der Physik* (Freiburg, 1965).

II

DISCUSSION OF "REALITY" AND "FREEDOM"

Freedom

If we expect theology to give this answer, it must first make plain its philosophical premises. It appears that a man who meets a miracle—and this is what is decisive for the believer—must overcome his tendency to react immediately to a worldly reality. This means that the reality of the miracle must first be discussed so that there can be progress from the mere establishment of the facts to a formulated expression of faith, a proclamation or kerygma deserving to be called language. The discussion of Bultmann, in fact the attempt to analyze the relationship between faith and understanding,[4] suggests to the theologian that language must be as open to the revelation of the hidden God as history is to happening. This makes it necessary to go beyond a revelation which is merely information—and takes the form of the judgment that Jesus is the revealer[5]—to understanding. This understanding is possible when the questions leading to it are asked within the context of an eschatological existence—which arises historically with Jesus of Nazareth. It must be a search not just for information, but to discover whether it is possible for eschatology and historical worldly reality to coexist.[6] This is the only way to avoid the formalism which is content to label things and leave it at that instead of becoming more and more aware of the world and God.[7]

That theology was caught up for so long with historical and comparative religious objections and did not bother with the problems of the physicist is apparent in the anger of Pascual

[4] R. Bultmann, "Das Problem der Hermeneutik," in *Glauben und Verstehen* II (Tübingen, 1952), pp. 211-35.

[5] R. Bultmann, *Theologie des Neuen Testaments* (Tübingen, 1953), pp. 298f., 444, 497; English trans.: *Theology of the New Testament*, 2 vols. (New York, Scribner).

[6] H. Ott, *Die Frage nach dem historischen Jesus und die Ontologie des Geschichte* (Zürich, 1960), pp. 21-3.

[7] H. Gadamer, *Wahrheit und Methode* (Tübingen, ²1695), pp. 145f., 383ff., 431f.

Jordan[8] and the fierce but just remarks about this lack made by Carl Friedrich von Weizsäcker.[9] Jordan tries to expound God's ways of working in the world with reference to the theory of gaps in the necessary indeterminism of quantum physics; von Weizsäcker asks theology a string of questions. Starting from the assertion that faith in science is the dominant religion of our time,[10] he thinks it necessary for two reasons to ask for mankind's sake whether a truly liberating religion is a possibility.[11] First, the demythologizing of the world through Jewish and Christian theism has freed the world of gods and now seems no longer able to preserve the idea for which it has been so freed, while not yet belonging to the God of love;[12] second, in this world, the absolute autonomy of man in the planning of his future raises the question of how this God can be reconciled with human freedom.[13] Von Weizsäcker does not give the answers to his first question, answers which since Jean Paul, Dietrich Bonhoeffer, Paul Tillich and Dorothee Sölle[14] have been thought of as post-theistic theology. This silence may astonish a well-educated Protestant but the second question is its justification.

After mock battles with the gap theory of quantum physics which works statistically, for the sake of methodological agreement, the basic assumption is again made that everything in the universe accessible to physics is ordered according to laws. Thus irrationality is excluded and it is possible to believe that the phenomenal world is ordered by rational laws and comprehen-

[8] P. Jordan, *Der Naturwissenschaftler vor der religiösen Frage* (Oldenburg-Hamburg, 1963), pp. 22, 156, 192ff.

[9] C. F. von Weizsäcker, "Schöpfung und Weltentstehung. Die Geschichte zweier Begriffe," in *Die Tragweite der Wissenschaft* I (Stuttgart, 1964), pp. 1ff., 173ff.

[10] *Ibid.*, pp. 3-9, 111.

[11] *Ibid.*, pp. 14-19.

[12] *Ibid.*, p. 93.

[13] *Ibid.*, pp. 11, 201ff.

[14] J. Paul, *Werke* II (Munich, [2]1959), pp. 266, 269; D. Bonhoeffer, *Ethik* (Munich, 1958), pp. 174-82; P. Tillich, *Der Mut zum Sein* (Stuttgart, [3]1958), pp. 28ff., 113ff.; D. Sölle, *Stellvertretung* (Stuttgart, [2]1965), pp. 131ff.

sible to reason.[15] It is well known that this assumption is made
by Einstein and his followers. Everyone is agreed again in this,
even the quantum physicists, but they are divided in the conclusions drawn from this assumption about man's freedom and
religion, and these conclusions thus become the questions asked
by physics of a theology of freedom.

Einstein's school, for all its Cartesian rationalism,[16] is aware
that there is no way from what is to what should be.[17] But its
view of nature leads, by a loose connection with a Leibnizian
preestablished harmony, to a pantheism, a cosmic religiosity
which becomes in the end mere amazement at the order of
nature.[18] Freedom is thus conceived—on the Cartesian supposition that thought is the ordering factor in man—in a sociolo-
gizing version of a critical and idealistic notion as the denial of
permanent dependence. Freedom already exists for Einstein
when human thought breaks free of the prejudices of authority
and this freedom is guaranteed by law.[19] Ernst Cassirer has
shown that the experience of reality which underlines this Ein-
steinian view of freedom is a fundamentally pluralistic one, and
this analysis reveals Einstein's idealistic view of truth. The attempt to impose a synthetic unity on the pluralism of the world
experienced as physics, sociology, aesthetics and religion[20] origi-
nates with transcendental philosophy, but is in fact only a variant

[15] A. Einstein, *Aus meinen späten Jahren* (Stuttgart, ²1953), pp. 18,
20, 23-35, 132, 158, 235, 261.
[16] A. Einstein, *Mein Weltbild* (Frankfurt-Berlin, 1964), pp. 115-27.
Criticism of the above: W. Heisenberg, *Prinzipielle Fragen der modernen
Physik* (Leipzig-Vienna, 1936), pp. 91-8; *idem, Wandlungen in den
Grundlagen der Naturwissenschaft* (Stuttgart, ⁹1959), pp. 61-79; N. Bohr,
Atomphysik und menschliche Erkenntnis (Braunschweig, 1958), pp. 32-
83; C. F. von Weizsäcker, *op. cit.*, pp. 201ff.; *idem, Zum Weltbild der
Physik* (Stuttgart, ⁷1958), p. 245.
[17] A. Einstein, *Aus meinen späten Jahren, op. cit.*, pp. 26, 132, 261.
[18] *Ibid.*, pp. 21, 262f.; *idem, Mein Weltbild, op. cit.*, pp. 18, 109, 151,
171.
[19] A. Einstein, *Freedom. Its Meaning* (New York, 1940).
[20] I. Kant, *Critique of Pure Reason*, p. 236; E. Cassirer, *Einstein's
Theory of Relativity Considered from the Epistemological Standpoint*
(Chicago, 1923), pp. 108-10, 120; *idem, Zur modernen Physik* (Darmstadt, 1957), pp. 188, 197-200, 256f.

of the rationalistic concept of the oneness of reason.[21] Because
neither truth nor other people are in opposition, the God of
religion becomes religiosity, and freedom becomes a planned
construction of the reason.

Reality

Hugo Dingler[22] has suggested that in this view of the world
the decisive attribute of the real, the essential, is lacking, and
reality becomes as boring as clockwork. Dingler's remarks in-
validate themselves as metaphysical statements upon the metaphysical relations between man and essential reality, because
this reality is deemed irrational, and there can be no argument
about it because it is beyond the reach of the rational planning
of man.[23]

Even David Bohm's[24] pluralistic structural analysis, which
again leans heavily on Einstein and de Broglie, remains in the
last resort bounded by a constructional horizon which is mort-
gaged to the idealist postulate of the unity of reason. Starting
from the insight that scientific method contains the tendency to
progress continually, he discusses how the views of reality of
classical physics and quantum physics can be reconciled. This
is through the relativity of the inexhaustible diversity of the
world corresponding to as many different scientific methods.
Through these, of course, only partial aspects of reality can be
approximately grasped, but the structures of the world as a
totality of limitless complexity can in this way be exactly described. With the help of this structural analysis, scientific research seeks an absolute in the context of which it may study
the relative in its inexhaustible diversity. Although we may value
Bohm's sketch because of its methodological purity and because

[21] K. Hübner, "Beitrage zur Philosophie der Physik," in *Phil. Rund-schau* 4 (Nov. 1953), pp. 18f. 25.

[22] H. Dingler, *Der Ergreifung des Wirklichen* (Munich, 1955), pp. 175-99.

[23] *Ibid.*, pp. 176-86.

[24] D. Bohm, *Causality and Chance in Modern Physics* (London, 1958), pp. 96-100, 170.

it is concerned with the desire of every modern physicist to re-discover some unified conception of physical reality, nevertheless Heisenberg's criticism is just; the "reality" which he is attempting to describe is once again only a rationally constructed ideological superstructure.[25]

It is of course true that physics can only be united by searching into the structures of physical reality. But even those who avoid mathematizing or logicizing construction end up with a strongly idealizing construction of the world. This is true of Suzanne Bachelard's phenomenological attempt at explanation,[26] and of Carl Friedrich von Weizsäcker's reflections upon a logic suited to modern physics,[27] and for Niels Bohr's principle of complementarity.[28] Bachelard not only seeks to make the ever more complex structures comprehensible to the mathematical physicist, so that the universal may become clear in the particular, but she also tries to give us knowledge of the structures so that after the mathematicians have done their worst, the pure empirical fact may be seen again. But when she conducts this transcendental reduction phenomenologically with numerous examples, the object of these descriptions is only clear when non-human reality is mathematically structured—that is, formulated by apodictic or mathematical knowledge. Von Weizsäcker also holds this view of reality; for him the complementary calculus of quantum physics is true logic, even though he is prepared to regard this logic merely as an ontological hypothesis. We are finally left with Niels Bohr's exploration of reality in the regions of biology, culture, art and what he calls faith. Of course under certain conditions the mathematical logical hypothesis may prove

[25] W. Heisenberg, "The Development of the Interpretation of the Quantum Theory," in *Niels Bohrs and the Development of Physics,* ed. W. Pauli (London, 1955), pp. 17f.

[26] S. Bachelard, *La conscience de rationalité. Etude phenomenologique sur la physique mathématique* (Paris, 1958), pp. 19, 29-35, 52, 80, 89, 191-210.

[27] C. F. von Weizsäcker, *Zum Weltbild der Physik* (Stuttgart, ⁷1958), pp. 208, 270, 299-301, 312f.

[28] N. Bohr, *Physique atomique et connaissance humaine* (Paris, 1961), pp. 5, 31-3, 79f.

practicable, but because the conditions are expressed in mathematical symbols, the question of fact only arises in those areas
to which these apply.

The point is raised in the work of Duhem at the turn of the
century.[29] Bloch's research into Newton and Karl Ulmer's work
on Galileo[30] disclose the same constructional hypotheses. Without going into the necessary corrections of detail in the corresponding bibliographical works, we find that, in the history of
physics since late Scholasticism, research was determined by the
constructional hypothesis and its more or less idealistic picture
of reality. It may nevertheless be useful here to give a brief
bibliography.[31]

More important than a detailed notice is the consideration of
the challenge to metaphysics and hence theology in Sir Arthur
Eddington's paraphrase of an Augustinian world picture: "We
have seen that even where science has made most progress, the
mind has only got back out of nature what the mind put into
nature. We have discovered a mysterious footprint on the banks
of the unknown river—and behold, it is our own." [32]

The physicists take up the cry of this questionable hominizing
of an idealistically conceived world picture. They either appeal,
on the grounds of the practicability of the knowledge of structures, simply to that critical rationalism proposed by Bavink

[29] P. Duhem, *La Théorie physique. Son objet, sa structure* (Paris,
1906), pp. 232ff., 274.

[30] L. Bloch, *La philosophie des Newton* (Paris, 1908); K. Ulmer, "Die
Wandlung des naturwissenschaftlichen Denkens zu Beginn der Neuzeit
bei Galilei," in *Symposion. Jahrbuch für Philosophie* II (Freiburg, 1959),
pp. 293-347.

[31] Cf. the comprehensive bibliography in A. C. Crombie, *Augustine to
Galileo* (Oxford, 1958) and the newer works not included in it: M. Bo-
nelli, *Mostra Anniversale di Galileo Galilei* (Florence, 1964); J. O.
Fleckenstein, *Scholastik, Barock und exakte Wissenschaften* (Einsiedeln,
1949); H. Lange, *Geschichte der Grundlagen der Physik I. II* (Freiburg,
1954 and 1961); O. Loretz, *Galilei und der Irrtum der Inquisition*
(Kevelaer, 1966); G. M. van Melsen, *Naturwissenschaft—Technik. Eine
philosphische Besinnung* (Cologne, 1964).

[32] Quoted in J. H. Jeans, *Physik und Philsophie* (Berlin, 1950), pp.
114f.

and Dessauer[33] and limit with March the construction of an objective world to microscopic physics,[34] or they acknowledge the passing of the one physics[35] and opt for Eddington's subjective selectionism,[36] for Jordan's empirical positivism,[37] or perhaps, with Heisenberg and von Weizsäcker, they search for the source of all knowledge forgotten by physics.[38] Even Erwin Nickel's attempt to reconcile materialism and spiritualism, and to show, through an analysis of the material, the preeminence of the spiritual, does not save the world, but sacrifices it to immateriality.[39]

Martin Heidegger gives the right reason for the fruitlessness of this attempt when he says that in physics, as an exact science, the understanding of the being of the existing things is not in a position to consider the essence of place, time, movement, force and mass as its own problem.[40] Anyone who reads von Weizsäcker's paper on "Speech as Information" will see how right Heidegger is.[41] Clearly, philosophers can only show that the natural scientists are responsible for the revelation of nature as the object of rational inquiry.[42] Metaphysicians nevertheless continue to believe that reason does not of itself compel a man to action,

[33] B. Bavink, *Ergebnisse und Probleme der Naturwissenschaften* (Leipzig, [8]1944), pp. 264f.; F. Dessauer, *Naturwissenschaftliches Erkennen. Beitrage zur Naturphilosophie* (Frankfurt, 1958), pp. 9, 18f., 211-14.

[34] A. March, *Das neue Denken der modernen Physik* (Hamburg, 1957), pp. 23f., 28f., 41, 97-9.

[35] E. Schrodinger, *Die Natur und die Griechen* (Hamburg, 1956).

[36] A. Eddington, *Philosophie der Naturwissenschaft* (Munich, 1949), pp. 37f., 134, 159.

[37] P. Jordan, *Der gescheiterte Aufstand* (Frankfurt, 1956), pp. 31-4.

[38] W. Heisenberg, *Wandlungen in den Grundlagen der Naturwissenschaft* (Stuttgart, [9]1959), pp. 87f.; C. F. von Weizsäcker, *Zum Weltbild der Physik* (Stuttgart, [7]1958), p. 174.

[39] E. Nickel, *Zugang zur Wirlichkeit. Existenzerhellung aus den transmateriellen Zusammenhängen* (Freiburg, 1963), pp. 25-36, 142-8, 211-20.

[40] M. Heidegger, *Das Wesen des Grundes* (Pfullingen, [3]1949), pp. 13f.

[41] C. F. von Weizsäcker, "Sprache als Information," in *Die Sprache*, ed. Bavarian Academy of Fine Arts (Darmstadt, 1959), pp. 48-53.

[42] M. Heidegger, *Vorträge und Aufsätze* (Pfullingen, [2]1959), pp. 24-35, 60-70, 84f.

but only reveals the truth, when the power of being itself, from which man with his reason and everything else is sprung, makes all this over to being.[43]

The revelation of the truth is thus in the power of being itself and not in the work of inquiring into existing things and their concepts[44] and thence it follows that after "being and time" we cannot write "time and being" because there is no way of getting from the concept of being to being. And so philosophers, in the face of the conceptual constructions of the physicists, are left merely with the faith—to use the words of Hölderlin—that "where danger is, there salvation arises".[45]

Salvation does not come from the sciences, because its factual content is not what they are equipped to cope with,[46] but it is (as Jacques Maritain, von Weizsäcker and Karl Rahner show[47]) self-evidently given and preliminary to the most formal logic, or lies (according to Maurice Blondel) at the root of action[48] or is present as preliminary to being (as Walter Strolz says with Heidegger) in history;[49] therefore, salvation of the power of being does not lie in these modes of human thought or behavior, but is preliminary to them, and not in the grasp of exact science. Bernhard Welte is right when he says: "The lines do not come together anywhere within this world's horizon. . . . Everything lives where it really lives . . . in hope, whose fulfillment remains with the unutterable." [50]

[43] G. Siewerth, *Metaphysik der Kindheit* (Einsiedeln, 1957), pp. 51f.
[44] M. Heidegger, *op. cit.*, p. 84.
[45] *Ibid.*, pp. 36, 43ff.
[46] *Ibid.*, pp. 66-70.
[47] J. Maritain, *Distinguer pour unir ou les degrés du savoir* (Paris, 1932); K. Rahner, *Geist in Welt. Zur Metaphysik der endlichen Erkenntnis bei Thomas von Aquin*, ed. J. B. Metz (Munich, ²1957), pp. 88f.
[48] M. Blondel, *L'Action* (Paris, ²1950), pp. 19, 132-6, 154, 345, 365, 380, 498; *idem, Lettre* (Paris, 1956), pp. 34-7.
[49] W. Strolz, *Der vergessene Ursprung* (Freiburg, 1959), pp. 128ff.
[50] B. Welte, "Das Heilige in der Welt," in *Freiburger Dies Universitatis* (1948/9), p. 178.

III

CONVERGENCE OF PHYSICS AND THEOLOGY?
DIALOGUE WITH THEOLOGIANS

If physicists ask the philosophers to reestablish the unity of their physics on a non-mathematized, idealistic conception of reality (after the converging of all modes of experience and comprehension), and if the philosophers whom they question, who are regarded on account of the recent history of philosophy and physics as theoreticians of knowledge—that is, as scientists for the explanation of being—give a negative answer, the philosopher's reference to hope and faith in the possible self-revelation of being will be existentially unacceptable to physicists.

This appears to be the reason why physicists from Einstein onward are inclined to draw religious parallels to their faith in the progress of physics. They simply no longer accept the philosopher's philosophical beliefs and turn to dialogue with the theologians whose statements of belief they still accept. This tendency is evident in the Gifford Lectures and the discussions of the Paulusgesellschaft. In fact, for these discussions theologians obviously are chosen who are not afraid to study physics and philosophy. When in these discussions the witness of faith as the ground of the theologian's hope is in question,[51] this is an opportunity for theological self-knowledge, because involved are questions which physics asks theology. If physics, as is clear from the above-mentioned literature in spite of its inaccessibility, is now ready to give up the 19th-century hypostasizing of the concept of nature, theology for its part could consider the transcendental philosophical reflections of the physicists and learn through a transcendental-theological, eschatological-ecclesiological explanation to give up the hypostasization of the concept of the Church. A discussion of the possibility and scope of this double demythologizing would really give theology a chance to reconcile the two mighty enemies, Church and world.

[51] J. Moltmann, "Hope without Faith: An Eschatological Humanism without God," in *Concilium* 16: *Is God Dead?* (1966), pp. 25-40; J. B. Metz, "Experientia spei," in *Diakonia* 1 (1966), pp. 186-91.

In preparation for this, as is mentioned in nn. 20 and 21 of the *Pastoral Constitution on the Church in the Modern World* of Vatican Council II, the Church should set up a research institute in which specialists in the natural and social sciences could consider, among other things, the questions of physics to theology. If in this institute, in the complete freedom which is the necessary condition for scientific dialogue, answers were found to these questions, this would benefit the Church who since the *Constitution on the Church* (nn. 1, 31, 38, 48) has regarded herself as the sign of unity and hope for all mankind.

PART III

DO-C DOCUMENTATION
CONCILIUM

Office of the Executive Secretary
Nijmegen, Netherlands

Death and Afterlife[1]

I n his *Le paysan de la Garonne,* a disillusioned Jacques Maritain said: "There are three things that an intelligent priest must on no account preach about today: the hereafter, the cross and sanctity." Then he goes on to ask whether the mass of Christians any longer *think* about these realities, the immortal soul or eternal life.[2] This underlines two facts. First, official theology seems to maintain a doctrinal silence on such points as a possible afterlife; second, the faithful themselves seem to demand something meaningful about these ultimate issues. Insofar as Holland is concerned, a recent inquiry has produced some scientifically responsible information about the situation.[3] According to this inquiry 70 percent of Dutch Catholics believe in an afterlife, 59 percent in a heaven, 39 percent in the existence of a devil and a hell, and 38 percent in the existence of a purgatory. The figures are somewhat higher

[1] We wish to acknowledge the assistance given by J. van Genderen, D. de Petter, A. Hulsbosch, J. B. Metz and P. Schoonenberg. A useful bibliography of recent studies may be found in P. Müller-Goldkuhle, *Die Eschatologie in der Dogmatik des 19. Jahrhunderts* (Essen, 1966); A. Ahlbrecht, *Tod und Unsterblichkeit in der evangelischen Theologie der Gegenwart* (Paderborn, 1964); Y. Congar, "Fins derniers," in *Rev. des Sc. Phil. et Théol.* 33 (1949), pp. 463ff.; L. Boros, *Mysterium Mortis—Der Mensch in der letzten Entscheidung* (Olten-Freiburg im B., 1962).

[2] J. Maritain, *Le paysan de la Garonne* (Paris, ²1966), p. 17.

[3] Attwood Statistics N.V., Enquête 66.02.70 (Rotterdam, 1966); this has appeared under the title of *God in Nederland* (Amsterdam, 1967).

with Protestants, of whom 98 percent believe in an afterlife, 97 percent in a heaven, 87 percent in a devil, 84 percent in a hell, but only 4 percent in a purgatory. Therefore, Maritain's question whether the mass of people any longer think about an afterlife is answered, at least for Holland, in the affirmative: most Christians by far believe in an afterlife.

The first statement of Maritain, about an intelligent priest owing it to his prestige to keep silent about these things, is difficult to check. The only thing we can do is to provide a survey of what theologians and those engaged in the sciences of the mind have said about these points.

The original outline of this volume included an article about death, but this had to be abandoned because of the author's illness. Yet, in an issue where evolution plays such a large part the question of *death and afterlife* can hardly be ignored, particularly since in the pattern of evolutionary thought death is easily deprived of its due seriousness by considering it as a kind of temporary imperfection which evolution will overcome. H. Dolch points in this same volume to a similar danger with regard to sin. Morever, from the pastoral point of view many an intelligent priest will probably welcome some positive orientation in a documented survey of what is constructive in modern thought about life and death.

For the sake of clarity we have organized this survey by concentrating on three lines of approach: the apocalyptic, the teleological and the prophetic. Of course, other classification breakdowns are possible and perhaps even more logical. But we have preferred this one because H. Cox has concentrated on these same three lines in his article on the expectation of the future in this volume. We shall begin, however, with a brief explanation of the three terms used here.

The *apocalyptic* expectation of the future is dominated by the conviction that the present is unsatisfactory and as such must be totally destroyed. It will be replaced by a future that is completely new, totally hidden and unpredictable, and brought in from the outside. Applied to mankind, this means that our life must be totally destroyed, that death finishes all, and that this life will be replaced by a new, unpredictable, heavenly existence through some outside intervention—for instance, a new creative intervention by God. It therefore has some similarity with the presentation of the apocalypse: the earthly Jerusalem disappears in a fire and from the clouds the heavenly Jerusalem descends in order to replace the earthly Jerusalem which has been destroyed. In this view there is no clear connection between life on earth and life in heaven.

The *teleological* expectation presupposes a final concrete goal which we can guess at because there are already more or less clear indications in this life, such as a general disposition or desire. In this view there is a causal connection between this life and that of the future. Applied to our theme, the situation is something like this: Man already possesses a soul which is immortal in itself. This soul is tuned in on an already existing aim (*telos*): the vision of God, heaven, or (if not pursued) the opposite. In any case, this aim exists regardless of the fact whether it is intentionally pursued or not. There is therefore a definite inner connection here between this life and that of the future and both are already real and knowable in principle.

The *prophetic* expectation is mainly characterized by the fact that the future is not yet a datum and cannot yet be known in principle. The word "prophetic" does not mean here that the future is anticipated but that it is announced to man in the name of God as the object of man's hope. In principle, however, this future lies in the hands of the human person who accepts this responsibility for his life. This future does not come from outside but is inwardly linked with the creative aspect of this life, and principally with the moral aspect of this creativity—in other words, this creativity as a moral task.

Although today the third kind of expectation seems to predominate, sometimes accompanied by irresponsible reactions against the other two, we do not mean to prejudge any of these three lines of approach, partly because all three rarely appear in an exclusive form.

We deliberately leave out the question of spiritualism for the simple reason that this does not represent an expectation of the future but is rather a dream-reality of actual life, a vague mirror of what is already present. For reasons of methodical limitation we also are not dealing with the Christian view of death as punishment for sin.

All three approaches to the mystery of death and afterlife can find support in the scriptures, so that none of the three can claim to be "scriptural" as its exclusive prerogative. One can also find support for all three in the pronouncements of the magisterium. By and large one may even say that everything said in the course of the centuries about death and afterlife can be reduced to these three lines of approach, inasmuch as (1) either death finishes all and then there is no genuine continuity at all; or (2) death marks a transition which can be seen as a continuity of what we know from life, but without its burdensome and painful aspects, and particularly without the transitoriness and disintegration which we experience now; or (3) death means a radical change, a new situation which differs totally from anything we have experienced.[4] As J. B. Metz has stated, it means "the final significance and fulfillment of our human existence as it is inaugurated and accomplished in the history of human freedom itself".[5]

I

THE APOCALYPTIC APPROACH

The various views we bring together under this heading hail, on the one side, from a positivist interpretation of revelation and, on the other, from a materialist concept of man. Both are prompted by a kind of protest against what is thought to be the Catholic view of immortality or against the primacy of the spirit over matter.

Among orthodox Protestant theologians such as Karl Barth,

[4] R. Beerling, "Denken over de dood," in *De Gids* (1966), p. 202.
[5] Cf. J. B. Metz, *Christliche Anthropozentrik* (Munich, 1963).

G. C. van Niftrik, P. Althaus and G. van der Leeuw,[6] there arose a sharp protest against immortality toward the end of World War II. According to them the Christian message was more than the preaching of a natural or even Platonic immortality. They maintained that there was a radical opposition between Greek thought and scriptural revelation, that scripture announces the resurrection and theology turned this into immortality. Like most protests, this, too, led to extremism: "Protestantism still lives for a large part outside that metaphysical dualism of the soul in which the primitive animism, present throughout humanity, has found its philosophical expression. Soul and body are then the two 'parts' of which man is composed. The soul is an immortal substance which is tied to the mortal body. Death then consists of a breaking of this bond, while the soul, because of its immortality, continues to exist. This is Roman Catholic dogma, and this it has to be since the Roman Catholic Church has dogmatized the philosophical heritage of antiquity." [7]

It is true that we are spiritually Semites and intellectually Greeks and we have become too much aware of our historicity to accept that everything within man is also truly human, but this does not necessarily mean that we yield to such a crude dualism. The quotation seems to imply that the Catholic view makes man with his type of immortality a rival of God who is solely immortal (1 Tim. 6, 16). It is understandable that these Protestant theologians emphasize that death is the end of all: "Ruthless destruction and total renewal is our view of death. And our identity is preserved throughout death in the love of God, in Christ, in the grace-filled judgment of God. There is within us nothing that, as such, ensures our identity here and hereafter, now and later. We are destined to die. But . . . our

[6] K. Barth, *Kirchliche Dogmatik* III/4 (Zurich, 1951), p. 678; P. Althaus, *Die letzten Dinge* ([5]1949); G. van der Leeuw, *Onsterfelijkheid of opstanding?* (Kampen, [4]1947); G. van Niftrik, *Kleine Dogmatiek* (Kampen, 1947).
[7] G. van Niftrik, *Zie, de mens—Beschrijving en verklaring van de Anthropologie van Karl Barth* (Kampen, 1951), pp. 281ff.

names are written in the book of life." [8] And van der Leeuw writes: "The faith says that we shall go down, totally and irrevocably; nothing will remain unless God works a new miracle of creation and makes us rise." [9] And all this in order to save the sovereignty of God.

Cullmann, who is closer to our point of view, speaks in a more moderate tone.[10] He is less preoccupied with the sovereignty of God than with that conviction which underpins his whole theology, namely the scriptural concept of time and the idea of *salvation history* connected with this concept. For Cullmann salvation is inextricably intertwined with the time element. The climax in this course of time, the "full measure", the *kairos,* has already been reached in God's closeness to man in the historical figure of Jesus of Nazareth. It seems to us that Cullmann is less worried about the immortality or possible afterlife of man than about the unimaginable character of time and its concomitant, eternity. On the one hand, Cullmann wants to let man's actual life pass over into a fuller life *via* his death, while, on the other, he will not speak of a total destruction because then the reality of man would land itself outside the scope of time and so outside the scope of salvation. He therefore opts for a kind of "in-between" situation which comes close to what the Old Testament means by "sheol", a kind of "slumber" condition which contains on the one hand the oppressive feeling of being stripped of everything and on the other the definite assurance that this situation, which is but transitory, cannot separate us from Christ. . . . But the reassuring certainty prevails because the decisive event has already taken place: death has been vanquished" (p. 50). This introduction of an "in-between" situation seems to us simply to postpone the problem. The issue is: What is the final situation? What makes this situation certain,

[8] Quoted in the excellent survey of A. Hulsbosch, *Is de ziel onsterfelijk?* ('s Hertogenbosch, 1960), p. 7.

[9] *Onsterfelijkheid of opstanding?*, p. 38.

[10] O. Cullmann, *Immortalité de l'âme ou Résurrection des morts* (Neuchâtel-Paris, 1956); cf. M. Cuminetti, "Immortalità dell'anima o risurrezione dei corpi," in *La Scuola Cattolica* XCIII, 2 (March-April, 1965), pp. 142-56.

and has it any connection with the past life? Cullmann's answer
to these questions seem to point to the opinions of Barth and
van der Leeuw quoted above: "Man, truly and totally dead, is
recalled to life by a fresh creative act of God. What happens
is unheard of, a miracle of creation. What happened before was
also overpowering: a life, created by God, has been destroyed"
(p. 25). And when he speaks of the seriousness of Jesus' death,
he says: "If life must arise out of this death, a new creative act
of God is necessary, an act which recalls to life the whole of
man, all that God has created and has been destroyed by death,
and not merely a part of man" (p. 24).

Trillhaas, in a very detailed study, has correctly pointed out
that all the arguments against immortality used by these authors
(and at one time also by himself) boomerang against the argument for the resurrection. It does not require much thought to
see that belief in the resurrection is not exactly served by at-
tacking immortality.

All these authors have one trait in common: man's actual
life is totally destroyed by death and man's eventual afterlife is
referred to a factor which lies outside man's scope.

The Marxist view seems to fall within this perspective and is
sometimes rightly called an inverted apocalyptic view. For here,
too, death is the definitive end of every human individual.[11]
Instead, something wholly new will appear. What we call "man"
is but a particular form in which nature expresses itself; it is
matter becoming conscious in this particular species. Matter is
the primary element, and man is but a special organization of
this matter which temporarily rises above the determinism of
matter through the process of becoming conscious. But this
process ends decisively with death, although the descending individuals continue it. Nature and man remain enemies; in every
death nature vanquishes man, even if man were to invent a
method of staying biologically alive.

Ultimately, the question does not concern this life but rather

---

[11] W. Trillhaas, "Einige Bemerkungen zur Idee der Unsterblichkeit," in
Neue Zeitschr. f. System. Theol. u. Religionsphil. (²1965), pp. 143-61.

man's interpretation of all that the Marxist calls "nature". This nature acts in history as something that is totally autonomous and self-sufficient; it obviously has nothing to do with a creator and still less with a preceding purpose. It is true that man escapes the control of nature up to a point by means of his consciousness through which he can control it and by his power to give a meaning to things; on the other hand, he will inevitably be vanquished by nature, though without the meaning he has given to things getting lost. The future lies with nature; this future is brought closer by succeeding generations, but not one of the preceding individual human beings has any share in it. Nature organizes itself into an economic and social whole, and one simply does not worry about a possible afterlife because human life remains strictly limited to its biological existence; death is essentially meaningless and there is no point in thinking about it. Death can in no way affect the meaning my life has poured into the world: "I can give meaning to my life by making it share in the universal purpose which the species has begun to work out since its origin, namely, in the building up of a fraternal society in which man has made radically sure of his dominion over his material environment. I can, no doubt, also follow my personal inclinations, relations and ambitions. But, in the degree in which I refuse to cooperate in this universal project, I reduce my own existence to the mediocre and to individualistic limitations without being able to escape from the formative influences of the social environment. And so I renounce once and for all any influence on historical progress. In this sense I am either saved or damned in this world, that is, in the sense of responding or not responding to the vocation of the whole of mankind, embodied in the struggle of the proletariat for the liberation of all mankind." [12]

[12] G. Mury, "Le marxiste devant la mort," in Vie Spir. (May, 1966), pp. 230-55; on the Catholic side, cf. G. Martelet, Victoire sur la mort (Lyons, 1966); C. Cases, "Entretien avec Ernesto de Martino sur la mort, l'apocalypse et la survie," in Esprit (March, 1966), pp. 370-7; F. Fortini, "Note conjointe sur la fin d'un homme et la fin du monde," in Esprit (March, 1966), pp. 378-82.

But the meaning of this new future is no clearer than what the apocalyptic literature calls the new earth and the new heaven: "The end of this project is but a convenient word used to indicate the objective participation in an enterprise which is not yet conscious of itself, that is, if by consciousness we mean scientific consciousness." [12a] It is in any case not a future for the committed individual human being; the individual is totally consumed in this project for a better future. Death is but an absurd limitation beyond which there lies no future for the individual. This is a consistent materialism: "It is precisely because man's nature is biological that death is fatal. Yet, this death does not make the human enterprise meaningless; it does, however, ruthlessly limit its scope, since it remains contained within this world. Even if the investigations of some research workers were crowned with success, even if future generations were endowed with an unlimited biological existence, the fact would remain that our present generation is totally dedicated to nothingness." [12b] Here the only way open to the human element is the courageous acceptance of the absurd fatality of death as a limit and to give up any hope of finding any other meaning in this death. Nevertheless, the question keeps on nagging us whether, in all this life which becomes aware of itself in history, there is not somewhere a human force capable of breaking through this absurdity.

This view of the individual human being in orthodox Marxism still leaves a whole field of humanity unexplored. Even within Marxism this conviction begins to emerge, as in Ernst Bloch's *Das Prinzip Hoffnung*. It would appear that both the apocalyptic theology of orthodox Protestantism which protested against immortality because of the resurrection, and the apocalyptic dogmatism of Marxism which objected to a possible afterlife of man because of alienation and the rejection of this earth, have had their day. For the former we referred to the well documented study of Trillhaas; for the latter we should note Bloch's words: "The inner essence of existence is not touched by death;

[12a] G. Mury, *loc. cit.*, p. 243.
[12b] *Ibid.*, p. 249.

if the possibilities of this kernel of existence really succeed, it reaches beyond death." [13] Both schools of thought are already pointing in the direction of a teleological expectation of the future.

II

THE TELEOLOGICAL APPROACH

This approach to the problem of death is mainly characterized by the attempt to link present life with a possible afterlife across the boundary of death.[14] These thinkers are aware of the fact that one can only speak of death and afterlife from the vantage point of the present. If we condense this attempt into three main propositions, we do not pretend to do justice to all the various opinions; we only stress the main elements of this line of thought. If we speak about death from the point of view of our present life, we must admit that, through Heidegger, the precariousness of this life as an "existence toward death" (*Sein zum Tode*) has penetrated into the consciousness of Western man. And yet, it must be pointed out that death is not in itself necessary (D. de Petter). Second, we see that in man's activity during life there develops something that endures and that maintains itself against the menace of death (K. Rahner). Third, death is not a fatality which can only be accepted as a brute fact; it can also be humanized and turned into a human act (L. Boros, P. Schoonenberg, C. Geffré).

Two points have to be made here before we go any further. First, this line of thought accepts that death is present through-

---

[13] J. Moltmann, *Theologie der Hoffnung* (Munich, [6]1966), pp. 313ff.; *idem*, "Messianismus und Marxismus," in *Kirche in der Zeit* 15 (1960), pp. 291-5; G. Sauter, *Zukunft und Verheissung* (Zurich-Stuttgart, 1960) which is a clear comparison of Moltmann and Bloch. What Moltmann calls "Hoffnung" is called "Zukunft und Verheissung" in Sauter.

[14] D. de Petter, *Begrip en werkelijkheid* (Hilversum, 1964), pp. 217-33, where the author takes up a position between Heidegger and Sartre without diminishing in any way the "precariousness" of human life.

out life and concerns man as a whole. This makes these considerations consistent with a new view of man, a renewed anthropology. Second, these views appear to be purely philosophical while they are in fact inspired by the Christian message about life.

Over against Sartre's view of death as something extraneous (we "also" die) and more in agreement with Heidegger's view of death as part of the price we pay for life throughout our existence, de Petter constantly stresses the precariousness of human life. This human life in its very essence is marked by death. Death cannot be minimized. It affects the whole of man, for it destroys man's bodily condition and man is man precisely because his spirit is intertwined with this bodily condition. Now, death means that the body is totally withdrawn from the spirit and the body is thus dehumanized. Is then the spirit also dehumanized?

In a certain sense the answer is affirmative. In and through the bodily condition the human spirit realizes itself. Although in itself essentially non-physical and therefore not subject to beginning and decay, the spirit is committed to the bodily condition. Modern anthropology is faced here with a difficulty which seems insoluble. On the one hand, the spirit is transcendent and not subject to beginning and decay, and cannot be affected by the disintegrating process of the body. On the other hand, one cannot imagine how the spirit can still have an opportunity of expressing itself after the body has been withdrawn by death. The living no longer have access to the dead because the dead have begun their absence from this world. Nor can one imagine that the spirit expresses itself or communicates with this world and its fellowmen in another way than through the bodily condition. The believer, however, can see here a glimmer in all this darkness which is the heart of the message of the Gospel, namely, the glorified bodily condition of Christ.

But is it possible to arrive here at something more concrete than Paul's thought about "being with Christ" which is then "far better" (Phil. 1, 23)? Indeed, but this implies then a genuine bodily and effective afterlife of Christ.[15] This is, therefore, essentially more than living on in the thoughts of others or as a source of inspiration for the Christian life of the believers. It is true that this, too, is a human way of being immortal, an anticipation of what we confess in the resurrection of Christ. But it would be unsatisfactory to reduce this confession of the depth and quality of life as proclaimed in the bodily resurrection of the Lord to such a living on in the memory of people. It is rather the continuation of the risen and glorified Lord's personal life in which the believer becomes involved precisely after death.

This continued life of the Lord is not a mere repetition of Jesus' earthly life without our earthly dimensions; nor is it a glorified existence for him alone in the form of an everlasting rest, because he is not a redeemer resting on his laurels but already a redeemer in the full, actual and present sense of the word.[16] If we can think of the bodily resurrection of the Lord as a possible way in which the "displaced" spirit can express itself and be operative, we can also better understand the sacraments as the *reliquia incarnationis* (the remains of the incarnation, what persists among us from the incarnation) and particularly the eucharistic presence which makes us familiar with the bodily presence of the Lord already in this present life. Through this sacramental bodily presence of the Lord we can also think of communion with the departed who remain part of the fullness of the Lord's body. But with all this there remains the difficulty of presentation: *How* does the human spirit find a way of expressing itself in the bodily condition of the risen Lord? We shall return to this point in the third section.

The possibility for the human spirit to be involved in the redemption of the world through the bodily condition of the living Lord implies another aspect; it is precisely here that the

[15] C. Geffré, "La résurrection ou la victoire de l'esprit," in *Vie Spir.* (April, 1963), pp. 382, 390-3; J. Kremer, *Das älteste Zeugnis von der Auferstehung Christi.* Stuttg. Bibl. Stud. 17 (Stuttgart, 1966).

[16] P. Schoonenberg, "Christus' Verlossingsdaad," in *Bijdragen* 27 (1966), p. 481.

spirit is clearly shown to participate in God's own life. This shows a depth and unity of life which transcends the individual life and is essentially more than a natural immortality.

The various implications we have mentioned in pursuance of de Petter's thesis can be found in a quotation which sums up what we have said so far: "The 'separated soul' is not a 'pure' spirit but the absurd situation of an essentially bodily human being; it is a real 'incarceration', a complete dislocation of the human person from his normal situation . . . at least, if we leave out redemption. For through this redemption everything changes for the separated spirit. There is therefore a radical difference between the miserable *natural* immortality of the human soul and *Christian* immortality. The first is implied in the human personality, and as such it marks the transcendence of the human spirit. But because this transcendence is essentially 'embodied', the natural immortality of the human spirit in the departed is an unnatural situation. Christian immortality, on the other hand, is an implication of communion in grace with the living God. . . . This redemption does not consist in a setting free *from* the body but rather in the human spirit's communion in God's own life." [17]

As a second characteristic feature of this teleological approach we referred to Rahner's point[18] that the continued life of the dead person must not be thought of as in straight linear continuity with the past life as if the coach were simply given a change of horses. Rahner's extensive contribution to the debate is perhaps best described as a view of death as a "dialectical reality", a reality which must be approached from two directions simultaneously, each clarifying the other. In discussing death he stresses a threefold mutual relationship: between time and

[17] E. Schillebeeckx, *Wereld en Kerk*. Theologische Peilingen III. De mens en zijn lichamelijkheid (Bilthoven, 1966), pp. 241-2.

[18] K. Rahner, "Das Leben der Toten," in *Schriften zur Theologie* IV (Zürich-Cologne, 1961), pp. 429-38; *idem*, "Das Aergernis des Todes," *loc. cit.* VII, pp. 141-5; *idem*, "Ueber das Christliche Sterben," *loc. cit.*, pp. 273-83 (1966); J. Nicolas, "A la jonction du temps et de l'éternité," in *Vie Spir.* (March 1963), pp. 298-311.

eternity, between limitation and freedom, and between passivity and activity.

That he considers the time-eternity relationship as most important is probably due to Heidegger's *Sein und Zeit* and his view of human existence as an "existence toward death". This leads in the first place to a correction of the view which looks on eternity as starting after this life so that the present life becomes a kind of probation period for an eternity which will meet us from outside: "Eternity arises *in* time as the real fruit of time and a fruit matured through time. Such an eternity does not arrive after the present life in order to carry on life but is rather an eternity which is set free from time and by the same token makes this time something transitory in order that the real and definitive thing can be done freely and without obstruction."[18a] This introduces a moral factor. The goodness or badness of human actions cannot be minimized in such a way that this goodness or badness can always be undone and a fresh start made in an unceasing circle of a time which constantly repeats itself. This would destroy the essential historicity of time. The darkness which stretches out behind death can therefore be a final night or the harbinger of an eternal dawn, an utter void or a total fulfillment, but never independent of man's actions.

Thus we reach the second relationship, between *necessity and freedom*. At the death of a loved person we have the impression that the definite contribution made during his life is destroyed; death seems to nullify his faithfulness and his love. This is the most brutal element in the experience of death, namely, that it throws a final boundary around man's greatest contribution in love, faithfulness and communion. This inevitable limitation, this cessation of freedom and decision, is one view. The other lies in the nature of human activity itself. Whenever man takes a genuine decision, makes a definitive choice, he transcends what is transitory and reaches freedom. He freely establishes something that is good (e.g., peace on this earth) which was

[18a] K. Rahner, *Schriften* IV, p. 432.

not there and which can never be considered as not done in the historicity of this world. This view is reinforced by the Gospel message about the lastingness of God's kingdom and the true freedom to which man is called in Christ: "Where man gathers himself and so, in self-possession, freely stakes his own personality, he does not indulge in a passing triviality but gathers up time in lasting validity. Such a time cannot be measured by the purely external experience of time, nor can it be identified with simple continuity, and still less with that purely temporal finality which finishes with our present life." [18b]

The third relationship is that between *passivity and activity*. Death as a purely biological phenomenon seems to be inflicted upon man from without, something over which he has no control. Over against this mere passivity we have one of the basic convictions of Christianity, namely, that Christ redeemed us through his death. Christ's death, therefore, is a human action with lasting validity. In this context Rahner frequently points to the fact that in the eucharist we constantly celebrate the Lord's death until he comes again. Christ seized hold of death and made it his own. Man, too, can turn this neutral outward death into a personal act by the conscious and free control of his whole life. In his death Christ created a freedom for himself by which he could continue to reach all. The character of this act lies precisely in the believing surrender to what presents itself as a bottomless abyss but becomes for the believer a promising reality. If not, death becomes simply a nervous clinging to what will escape us in any case. Ultimately, we are "disposed of", not as a "thing", however, but as a person who is set free. The *act* of dying consists in "the continued belief in this 'dispensation' ": "It is the willing acceptance of what is decided, and our adjustment to it; it is the anticipated abandoning of what we shall have to renounce in any case in the belief that this poverty, this acceptance of what is decided about us, sets us free for the blessed and immeasurable goodness of the deciding

[18b] *Ibid.*, p. 434.

God. . . . If we do not accept this, death can only be an un-spoken protest against death in this life." [18c]

This last view already touches on what we have called the third form of teleological expectation, namely, that actual death is not a mere fatality but something that can be humanized and turned into a human act.[19]

The authors dealt with so far shift the real problem of death to what precedes the biological death, in contrast to earlier treatments which looked more toward the realities on the other side: heaven, hell and purgatory. The next group of authors concentrate more on the act of dying itself, not in what precedes it or follows it, but on the moment of transition, and this coincides with the moment of death which cannot be expressed in purely temporal terms. This approach has more the character of a "working hypothesis" and its probability or lack of it depends on the harmonizing of as many data of faith as possible.

Without delving into philosophical presuppositions, the essence of this theological hypothesis can be summed up as follows. Every dying person makes a final choice for or against God in a last, truly free act of his present life. This act also contains the judgment because this act takes into account the whole of the preceding life. There is no experimental certainty about the existence of this act. It is a kind of ultimate summary of all one's human decisions, a kind of synthesis of all those elements in our human actions which, however obscurely, contain something of a turning away from or toward God. It is the decisive last step on a whole pilgrimage which leads man to his final goal or to the end of a dead alley. It brings out the basic option

[18c] *Ibid.*, VII, p. 278.
[19] P. Glorieux, "Endurcissement final et grâces dernières," in *Nouv. Rev. Théol.* 59 (1932), pp. 865-92; P. Schoonenberg, "Onze activiteit in het sterven," in *Bijdragen* 6 (1943-5), pp. 127-45; E. Hengstenberg, *Einsamkeit und Tod* and *Tod und Vollendung* (both Regensburg, 1938); A. Winklhofer, *Das Kommen seines Reiches* (Frankfurt a. M., 1959); R. Troisfontaines, *Je ne meurs pas* (Paris, ⁵1960); idem, *J'entre dans la vie* (Paris, ⁶1963); L. Boros, *Mysterium mortis. Der Mensch in der letzten Entscheidung* (Freiburg im B., 1962); C. Geffré, "La mort comme nécessité et comme liberté," in *Vie Spir.* (March, 1963), pp. 264-80.

that has influenced our whole life. Thus this view bridges the gap between time and eternity and saves both the continuity of human life and the identity of the person. L. Boros[19a] even tries to link the final situation of the unbaptized children with this "decision hypothesis".

Modern man experiences here a special difficulty that provides the answer for the secondary questions but simply takes the afterlife for granted. But modern man is in doubt precisely about this privileged condition which he would occupy in the whole sphere of living beings. He is too well aware of the laws of biology and the oneness of matter to accept without more ado a human prerogative of continuity without being subject to the general law of disintegration. But here Teilhard de Chardin's vision of an evolution which is still necessary to human life in order to reach a higher level opens up some possibilities. One must credit H. de Lubac with having pointed this out.[20]

All these authors have in common the fact that, in various ways, they all stress the anthropological dignity of man. They all have helped to free the preaching of life and death from a suspected alienation of man from himself. They all take as a basic fact that there is a certain continuity between this life and afterlife.

III

THE PROPHETIC APPROACH

If we were to say that the previous approach was basically oriented toward a beatific *vision,* we might say of the prophetic approach that it is concerned with a *spes beata,* a beatific hope. In this case we must be able to present the realities of life and death as desirable. The appeal of such values depends on their possible realization.

[19a] L. Boros, *op. cit.,* pp. 116-22.
[20] H. de Lubac, *La pensée religieuse du Père Teilhard de Chardin* and *La Prière du Père Teilhard de Chardin* (Paris, 1964), pp. 121, 149-64.

As soon as we mention the words "desire" and "appeal", the shades of Freud rise up to warn us that religion is an illusion and that heaven and hell are but the projection of a desire.[21] This then leads to an easy interpretation of the reality: man's desire for happiness and afterlife is reduced to an instinct, the instinct is identified with what is primitive, and what is primitive must be rejected as not belonging to the evolutionary phase of awareness and historicity, Afterlife becomes a primitive myth, and heaven and hell are part of a language that cannot be verified. But is it not possible that myths could set us thinking? Is there not room here for what Geffré has called a new human self-interrogation at the very heart of the faith? [22]

What strikes us here first of all is that the problem of life and death is no longer a purely religious problem but a universal one with which psychoanalysis, linguistic analysis and hermeneutics are concerned, and not merely as with a border problem.

What Freud particularly has taught us about our urges, particularly the life wish and the death wish, has made us aware of the fact that the human desire for continuity or destruction, expressed in the words "heaven" and "hell", contains a large element of illusion. The question which both the philosopher and the theologian have to ask themselves is: How far can we eliminate this illusion? What Freud has said about desire and Jung about the archetypes does not exhaust the problem. This diminution of idealistic expectations and desires as illusions was necessary, but it only stresses the downward view. Paul Ricoeur has tried to show that there is also room for an "upward" view. And now the theologian has the burdensome task to point courageously to the illusive element in the Christian expectation without destroying the expectation itself. There is no room here to refer to the vast amount of practical work done already in the

[21] P. Ricoeur, *De l'interprétation* (Paris, 1965); J. Pohier, "Au nom du Père," in *Esprit* (March and April, 1966), pp. 480-500, 947-71, outlines the work to be done by theology in this field.

[22] C. Geffré, "La recherche en théologie," in *Vie Spir.* (Feb., 1967), p. 21.

new catechetical treatment of this expectation. There is, however, a danger that, for the sake of psychological hygiene, death
and life will no longer be taken seriously when we explain away
the ultimate issues. We must develop a new understanding of
the reality in its symbols.

The work done by *analytical theology,* particularly in Anglo-
Saxon countries, has been very helpful here. For the language
which speaks about the mystery of death and life is not an exact
language. The question is, however, whether scientific and exact
language is the one and only language. It may be that a language
of persuasion shows another structure than scientific language.
And here lies the merit of such people as Wittgenstein. It is not
that he can straightway save the language about life and death
from its present difficulties, but he makes the theologian aware
of the fact that a language of persuasion uses other criteria than
scientific language. Any speech from a specific angle implies inevitably a methodical limitation of that speech. This does not
mean that we can make no statements which are absolutely
valid or true, but rather that these statements are only true
within a given game of language, which follows its own rules.
We can never speak absolutely about the absolute, but we can
speak of it from a particular angle. If we use the term "game",
the intention is not that it should not be taken seriously. On the
contrary, there is a vast quantity of linguistic games, and theology also can be looked at as a collection of linguistic games.
Every such game has a coordinating principle which allows us to
obtain the relatively best coordination of elementary statements.

These technical details had to be mentioned in a general way
in order to prevent us from considering the prophetic approach
as altogether too subjective. The coordinating principle for
these analytical theologians is *hope.* If we start with "hope", it
is obvious that we shall develop a linguistic "game" which has
different rules from that of the natural sciences. To say that
"human life is indestructible" is meaningless in the linguistic
game of the natural sciences but not in that of hope. In this

field it is worth our while to listen to S. Ogden.[28] This author of *Christ without Myth* approaches the matter of life and death from the angle of *trust,* the trust in God's love for us. On this trust he builds up an argument which discusses the realities that the now meaningless myths and symbols aimed at. He disagrees essentially with authors like W. Hamilton,[24] T. Altizer and P. van Buren, who demand a radical secularization also in this matter. He follows Wittgenstein[25] in not wanting to increase the puzzles of life by talking about an "everlasting" life but in trying to find something that ensures a way of overcoming the "perpetual perishing". The strength of these authors lies not in one or other particular statement but in the "discourse" as a whole which is correlated to our human existence but is never wholly enclosed by it.

To show how such a "discourse" works, it is best to start from Ogden's view of God's love. To love means to put the center of my existence in someone else. Does God not do precisely this in Jesus Christ? God becomes totally God insofar as he puts the center of his concern in Jesus and through Jesus in each of us. What this means in the concrete I learn constantly afresh in the way I adjust my life to the Gospel. The obvious meaning of the narrative of the resurrection is precisely that God does not leave this Jesus in the disillusion of death but remains God for him behind the frontier of death. Jesus remains the center of God's redeeming concern.

Can this be a message for my life in this age? Through my human experience I have access to the life-giving element which I find in someone's loving concern for me. I know of situations in which I am silenced to death but I also know of situations in which I become myself through the concern of another. And

[23] S. Ogden, *The Reality of God* (London, 1967), pp. 206-30; *idem, The Promise of Faith.* Cf. also F. Ferre, *Language, Logic and God* (New York, 1961), p. 123.

[24] W. Hamilton, *Till Death Us Do Part* and *Room for Death* (St. Louis, 1964).

[25] L. Wittgenstein, *Tractatus Logico-Philosophicus* (London, 1922), p. 184.

here I find a motive to trust myself to the same assuring reality which did not leave Jesus in the state of death. It is therefore possible to find the ultimate meaning for my life in the Gospel. And yet, this message about my life is not primarily a communication which leads to self-understanding but rather a communication about the nature of God who is love, and so stronger than death. It is theology.

Obviously, even this does not provide a final solution for the problem of life and death. But it does create an opportunity for formulating the questions differently. Bonhoeffer[26] once raised the poignant question of what would happen to belief in God when the ultimate questions can also be answered without God. With van Iersel [27] one might reply: *"Then the last possibility of misunderstanding God will have disappeared and our belief in God's faithfulness might be purer beyond the grave than it is now."* And is this not one of the enduring problems of all theology?

This last sector has not dealt with the more explicit treatment based on hermeneutics[28] or with the relations between sexuality and death.[29] Nevertheless, whoever has been able to assimilate what we have said in this too brief survey will hardly complain with Maritain that not enough thought is given to death among contemporary theologians. He will, however, notice that the attention has shifted from "after death" to "life" and from imprisonment to freedom. But this seems to us rather a gain.

[26] D. Bonhoeffer, *Widerstand und Ergebung.* Briefe und Aufzeichnungen aus der Haft (Munich, 1951), pp. 159-60.

[27] B. van Iersel, "Vragen naar dood en leven in het Nieuwe Testament," in *Verbum* (May, 1966), p. 196.

[28] Here one should carefully examine P. Ricoeur, *Le volontaire et l'involontaire* (Paris, 1963), pp. 428-35; *idem, Finitude et culpabilité, II: La symbolique du mal,* pp. 243-60, 264-70.

[29] M. Oraison, "A propos d'une théologie de la mort," in *Vie Spir.* (May, 1966), pp. 215-7.

BIOGRAPHICAL NOTES

WERNER BRÖKER: Born in Recklinghausen, Germany, in 1929, he was ordained in 1955. He studied at the Universities of Münster, Munich and Bonn, and received doctorates in the natural sciences (1962) and in theology (1967). He is scientific assistant at the University of Münster, and has published *Sinn der Evolution* (Düsseldorf, 1967).

ZOLTÁN ALSZEGHY, S.J.: Born in Budapest in 1915, he was ordained in 1942. He studied at the Gregorian University, Rome, where he obtained his doctorate in theology in 1945, and where he is at present teaching. Among his published works are *Il Creatore* (Florence, 1959) and *Il Vangelo della grazia* (Florence, 1964), both in collaboration with M. Flick, S.J.

HARVEY COX: Born in 1929 in the United States, he was ordained a minister of the Baptist Church in 1956. A graduate of Yale and Harvard Universities, he obtained his doctorate in philosophy in 1963. Until 1965 he was assistant professor of theology and culture at Andover Newton Theological School, and is now theological professor of the Divinity School at Harvard University. He has published *The Secular City* (1965) and *God's Revolution* (1965) and was one of the editors of *Christianity and Crisis*.

ANDREAS VAN MELSEN: Born in 1912 at Zeist, Netherlands, he studied at the University of Utrecht. He gained the degree of Doctor of Science in 1941. Since 1945 he has been professor of the philosophy of science at Nijmegen University, and since 1953 he has been honorary professor in the same subject at the University of Groningen. He is dean of the faculty of mathematics and science at Nijmegen University, and president of the Committee of the Dutch Pastoral Council. Among his published works is *Evolution and Philosophy* (Pittsburgh, 1965).

KARL RAHNER, S.J.: Born in Freiburg im Breisgau in 1904, he was ordained in 1932. He studied at the Universities of Freiburg im Breisgau and Innsbruck, receiving his doctorate in theology in 1936. He was formerly professor of the philosophy of religion and Christian anthropology at Munich University, and at present is professor of dogma and the history of dogma at the University of Münster. Among his many important works are *Schriften zur Theologie* (tr.: *Theological Investigations,* London & Baltimore) in seven volumes (1954-1966). He is also editor of the ten-volume *Lexikon für Theologie und Kirche* (Freiburg im Breisgau, 1957-1965) and of *Handbuch der Pastoral-theologie,* Vols. I-II (Freiburg im Breisgau, 1964-66).

HEIMO DOLCH: Born in 1912 at Böhlitz-Ehrenberg, Germany, he was ordained in 1946. He studied at the Universities of Leipzig and Louvain, receiving doctorates in philosophy (1936) and in theology (1951). He has been a professor at Bonn University since 1963. Among his publications are *Theologie und Physik* (1951) and *Teilhard de Chardin im Disput* (1964).

DOMINIQUE DUBARLE, O.P.: Born in 1907 in France, he was ordained in 1931. He became a doctor of philosophy and theology in 1933, and since 1944 has been professor of philosophy at the Catholic Institute, Paris. He has published *Pour un dialogue avec le marxisme* (Paris, 1964).

JACQUES ELLUL: Born in Bordeaux in 1912, he is a member of the Reformed Church of France. He studied at the University of Bordeaux and the Faculty of Law in Paris. Admitted to the Bar in 1943, he is professor of law at the Institute of Political Studies in the University of Bordeaux. In addition, he is a member of the Ecumenical Commission of the Reformed Church of France and also of its National Council. Among his publications is *Politique de Dieu, Politiques de l'homme* (1966).

EMMANUEL MESTHENE: Born in the United States in 1920, he studied at Columbia University, where he gained an M.A. in 1949 and his doctorate in philosophy in 1964. In the same year he became executive director of the Program on Technology and Society at Harvard, where he is also in charge of the Graduate School of Business Administration. He has published *The Meaning of Technology for Man* (1967).

ERIC MASCALL: Born in London in 1905, he was ordained in the Anglican Church in 1932. He studied at the Universities of Cambridge, Oxford and London, receiving an M.A. in mathematics and gaining his doctorate in theology in 1948. He has been professor of historical theology at King College, London University, since 1962. He is the author of *The Secularisation of Christianity* (London & New York, 1965).

BEN VAN ONNA: Born in The Netherlands in 1940, he has studied at the University of Münster and is preparing for his doctorate in theology at the University of Tübingen under the direction of Professor Joseph Ratzinger.

NORBERT SCHIFFERS: Born in 1927 at Aix-la-Chapelle, he was ordained in 1952. He studied at the University of Tübingen, receiving his doctorate in theology in 1954. He is head of studies at the University of Münster, where he teaches fundamental theology. Among his published works is *Die Einheit der Kirche nach J. H. Newman* (Düsseldorf, 1956).

International Publishers of CONCILIUM

ENGLISH EDITION
Paulist Press
Glen Rock, N. J., U.S.A.

Burns & Oates Ltd.
25 Ashley Place
London, S.W.1

DUTCH EDITION
Uitgeverij Paul Brand, N. V.
Hilversum, Netherlands

FRENCH EDITION
Maison Mame
Tours/Paris, France

GERMAN EDITION
Verlagsanstalt Benziger & Co., A.G.
Einsiedeln, Switzerland

Matthias Grunewald-Verlag
Mainz, W. Germany

SPANISH EDITION
Ediciones Guadarrama
Madrid, Spain

PORTUGUESE EDITION
Livraria Morais Editora, Ltda.
Lisbon, Portugal

ITALIAN EDITION
Editrice Queriniana
Brescia, Italy